Statistics & Computing

W. Härdle · L. Simar

Editors

Computer Intensive Methods in Statistics

Springer-Verlag
Berlin Heidelberg GmbH

Statistics and Computing

Wolfgang Härdle
Léopold Simar (Eds.)

Computer Intensive Methods in Statistics

With 34 Figures

Springer-Verlag Berlin Heidelberg GmbH

Prof. Dr. Wolfgang Härdle
Institut für Statistik und Ökonometrie
Wirtschaftswissenschaftliche Fakultät
Humboldt-Universität zu Berlin
Spandauer Straße 1
O-1020 Berlin, FRG

Prof. Léopold Simar
Institute de Statistique
Université Catholique de Louvain
B-1348 Louvain-la-Neuve, Belgium

ISBN 978-3-7908-0677-9   ISBN 978-3-642-52468-4 (eBook)
DOI 10.1007/978-3-642-52468-4

Preface

The computer has created new fields in statistic. Numerical and statistical problems that were untackable five to ten years ago can now be computed even on portable personal computers. A computer intensive task is for example the numerical calculation of posterior distributions in Bayesian analysis. The Bootstrap and image analysis are two other fields spawned by the almost unlimited computing power. It is not only the computing power through that has revolutionized statistics, the graphical interactiveness on modern statistical environments has given us the possibility for deeper insight into our data.

On November 21,22 1991 a conference on computer Intensive Methods in Statistics has been organized at the Université Catholique de Louvain, Louvain-La-Neuve, Belgium. The organizers were Jan Beirlant (Katholieke Universiteit Leuven), Wolfgang Härdle (Humboldt-Universität zu Berlin) and Léopold Simar (Université Catholique de Louvain and Facultés Universitaires Saint-Louis). The meeting was the XIIth in the series of the Rencontre Franco-Belge des Statisticiens. Following this tradition both theoretical statistical results and practical contributions of this active field of statistical research were presented. The four topics that have been treated in more detail were: Bayesian Computing; Interfacing Statistics and Computers; Image Analysis; Resampling Methods. Selected and refereed papers have been edited and collected for this book.

1) Bayesian Computing.

In Bayesian analysis, the statistician is very often confronted with the problem of computing numerically posterior or predictive distributions. In the field of image analysis D.A. STEPHENS and A.F.M. SMITH show how the Bayesian idea of Gibbs sampling helps to solve computational problems in Edge-Detection. F. KLEIBERGEN and H.K. van DIJK analyse how the generation of matric variate t drawings may be used to approximate the computation of posteriors in econometric problems. L. BAUWENS and A. RASQUERO consider high posterior density regions to solve the problem of testing residual autocorrelation.

2) Interfacing Statistics and Computers.

D. WOUTERS and L. VERMEIRE use the NAG library for the numerical construction of optimal critical regions in the gamma family and show how the Mathematica language can be used for the symbolic computation of a geodesic test in the Weibull family. A. ANTONIADES, J. BERRUYER and R. CARMONA report on teachware experiments centered around the use of a dedicated program in the learning of mathematical statistics and the practice of data analysis. In this program, graphical tools, random number generation, simulation techniques and bootstrap are integrated with the idea of guide undergraduates to modern statistical techniques.

3) Image Analysis.

In M. ROUSSIGNOL, V. JOUANE, M. MENVIELLE and P. TARITS a method for finding rock conductivities from electromagnetic measurements is presented. The method uses a stochastic algorithm to find a Bayesian estimator of the conductivities. Markov random field models are used in V. GRANVILLE and J.P. RASSON in image remote sensing. The technique is extended, in a unifying approach, in order to analyse problems of image segmentation, noise filtering and discriminant analysis. In A.P. KOROSTELEV and A.B. TSYBAKOV minimax linewise algorithms for image reconstruction are studied.

4) Resampling Methods.

Ph. VIEU presents a survey on theoretical results for bandwidth selection for kernel regression. The behavior of kernel estimates of a regression function depends heavily on the smoothing parameter, so for practical purpose it is important to ensure good properties of the estimates. M.A. GRUET, S. HUET and E. JOLIVET consider the bootstrap in regression problems. Four specific problems are considered: confidence intervals for parameters, calibration analysis in nonlinear situations, estimation of the covariance matrix of the estimates and confidence intervals for the regression function. Finally, Ph. BESSE and A. de FALGUEROLLES investigate the use of resampling methods in data analysis problems. More specifically, they study the problem of the choice of the dimension in principal components analysis. Different bootstrap and jackknife estimates are presented.

Several institutions have made the XIIth Franco-Begian Meeting of Statisticians possible; their financial help is gratefully acknowledged: CORE, the Center for Operation Research and Econometrics (Université Catholique de Louvain), SMASH, Séminaire de Mathématiques Appliquées aux Sciences Humaines (Facultés Universitaires Saint-Louis), KUL, Katholieke Universiteit Leuven, the Ministère de l'Education de la Communauté Francaise de Belgique and the FRNS, Fonds National de la Recherche Scientifique, all helped to make this meeting possible. We would like to thank all them. The organisation would not have been possible without the staff of CORE, in particular Sheila Verkaeren and Mariette Huysentruit. We cordially like to thank them.

W. Härdle   L. Simar

Contents

Bayesian Edge-Detection in Images via Changepoint Methods

D. A. Stephens and A. F. M. Smith

Department of Mathematics, Imperial College London.

180 Queen's Gate, London SW7 2AZ, United Kingdom.

Abstract

The problem of edge-detection in images will be formulated as a statistical changepoint problem using a Bayesian approach. It will be shown that the Gibbs sampler provides an effective procedure for the required Bayesian calculations. The use of the method for "quick and dirty" image segmentation will be illustrated.

Keywords : Image analysis; edge-detection; changepoint identification; Bayesian statistics; Gibbs sampler; edge reconstruction; image reconstruction.

1. Introduction

1.1 Background

Practical applications of image analysis abound in agronomy (remote sensing), astronomy (study of galaxies), industrial processing (automated manufacturing and quality control), medicine (internal body imaging), and the military (intelligence, reconnaissance, defence/offence systems), relating variously to imaging technologies such as photography, tomography, radiography etc. In this paper, we shall discuss a "quick and dirty" statistical approach to the problem of edge-detection in such images.

We shall denote the observed data image by $\mathbf{Y}$, and true scene by θ. Generally, $\mathbf{Y}$ and θ will be vectors of possibly different lengths, reflecting discretization involved in the image formation process. It is possible, in specific problems, to regard the true scene as having a continuous parameter space, but here we restrict ourselves to the discretized version. For definiteness, we shall focus in what follows on the situation where $\mathbf{Y}$ and θ have equal length, M, and relate to the same rectangular grid. Our interpretation of the image formation process is a conventional statistical one, namely that

$$Data = Structure * Noise ,$$ (1)

where the terms "*Data*" and "*Structure*" in (1) correspond, respectively, to "image" and "true scene" as defined above, "*Noise*" corresponds to the inherent but undesirable stochastic element, and * is an operator defining precisely how the *Structure* and *Noise* interact. In common terminology, a noise-process is regarded as acting to corrupt the underlying signal. Hence, we denote the noise-process by ε, and thus we may formally think of (1) in the image processing context as

$$Y = f(\theta, \varepsilon) \tag{2}$$

where f is merely some function involving the operation * and the assumed Y, θ pixel correspondence described above.

We shall adopt a Bayesian approach to inference about aspects of the unobservable θ of interest (in our case edges). First, we provide a general background and motivation.

1.2 Problems in image processing

In image segmentation, given the observed (image) data Y, the objective is to allocate or classify each of the elements in the unobservable true scene vector θ, to one of a set of "textures". The segmentation problem may be approached in several ways, such as estimation, probabilistic classification etc., and is often the prime objective in many fields of application, where the unobservable true scene is thought to comprise of broadly homogeneous texture regions separated by thin boundaries or edges.

It is often of interest to be able to identify the positions of these edges, either as an end in itself, or as a preliminary stage in segmentation. We note below that, despite the considerable literature concerned with edge-detection methodology and applications, little has been done to formulate the problem in a formal probabilistic framework. This latter task is the primary concern of this paper.

In many problems, for example in medical imaging or military reconnaissance, the inference problem involves discovery of the location of a high intensity object of small dimension relative to the background image field. This problem of object detection is a somewhat specialized version of the segmentation problem described above. Although edge-detection is also relevant here, we have the additional knowledge that edges of the object are "close together".

A pattern is a particular configuration of texture regions or features in the true scene, having a characteristic quality that allows discrimination between it and other configurations. Again, edge detection has at least a preliminary role to play in many pattern recognition applications.

1.3 Statistical approaches to image segmentation

Image segmentation has been approached in a statistical framework in several ways. First, it has been viewed as an estimation problem, where the elements of θ, the true scene pixel classification values, are regarded as unknown parameters that may be estimated using classical (maximum-likelihood) or Bayesian (maximum probability) techniques. Secondly, it has been viewed as what we shall call a probabilistic classification problem, approached via such procedures as cluster analysis, discriminant analysis, and predictive classification, where the elements of θ are allocated to textures according to their fidelity to texture characteristics. Other, less formally probabilistic, classification techniques have also been proposed. We concentrate here solely on the Bayesian estimative approach.

If we denote the probability distribution for Y given θ by $[Y \mid \theta]$, then, in a classical statistical framework, inference about θ is made, for example, via maximum-likelihood methods. In a Bayesian decision-theoretic framework, inferences are made via the posterior distribution for θ given Y, denoted by $[\theta \mid Y]$. In this framework, the required estimate for θ is derived from the joint posterior distribution, possibly the joint posterior mode (M.A.P), or the modes of the marginal posterior distributions for the elements $\theta_1, \cdots, \theta_M$ of θ (M.P.M). Throughout, we assume that distributions have density or mass function representations.

Clearly, before we are able to report the required estimates we must first evaluate the joint posterior distribution, $[\theta \mid Y]$, or the set of marginal posterior distributions, $[\theta_i \mid Y]$ for $i = 1, \cdots, M$. In the image segmentation context, for definiteness, we shall consider the specification of a joint prior distributions for the true scene parameters, denoted by $[\theta]$, and the evaluation of the joint posterior distribution $[\theta \mid Y]$. It is clear that the joint and marginal prior and posterior distributions are, in fact, deterministically related, and that specification of a joint structure induces a marginal structure. Via Bayes theorem we have (up to proportionality, ignoring the normalizing constant)

$$[\theta \mid Y] \propto [Y \mid \theta][\theta], \tag{3}$$

where the first term on the right-hand side can be any version of the likelihood function derived from the distribution $[Y \mid \theta]$. The general problem thus reduces to specifying appropriate forms of likelihood and prior, and identifying and evaluating the posterior distributional form that appears in (3).

If beliefs about the true scene are that it comprises homogeneous texture regions separated by edges, the latter being regarded as small-scale features relative to the size of the texture regions, then in any localized sub-region of the true scene we would generally expect contiguous blocks of pixels of the same texture to exist, with isolated pixels of any texture rarely occurring. One possible specification of such beliefs is via a Markov Random Field. Such prior distributions have frequently been used to model the true scene in statistical image segmentation procedures, and summarization and maximization of the resulting posterior distribution is then feasible using the Gibbs sampling algorithm (Geman and Geman (1984)) to which we shall return later with a somewhat different emphasis.

4

1.4 Edge-detection

Edge-detection is an important preliminary operation in many forms of image analysis and the problem has received considerable attention in the classical image processing literature. A review of edge-detection techniques can be found in Rosenfeld and Kak (1982, chapter 10). A variety of techniques based on localized differential operators under simple statistical (Gaussian) assumptions for the image model have been proposed. Such techniques commonly involve numerical differentiation or approximation to differentiation, and subsequent optimization of the first derivative (extrema methods), or location of positions where the second derivative is zero (zero-crossings methods). See, for example, Torre and Poggio (1986) for a discussion of the two related (but non-equivalent) techniques and a description of a regularization approach to the edge-detection problem, and the work of Canny (1986), who also uses an essentially regularization-based approach. As such techniques commonly revolve around local operations performed in series for different sub-images or "windows" within the complete image, subjective choices must be made concerning the size of window used and the precise way in which the results from different windows of possibly different sizes are to be combined: see Lu and Jain (1989) for a discussion of such problems. However, we are of the opinion that the localized nature of such techniques conflicts with our global interpretation of many edge-detection problems. We later present a re-formulation of the edge-detection problem which utilizes more directly the global information from the observed image.

Edge-detection techniques of a rather more formal statistical nature have also been proposed. Mascarenhas and Prado (1980) devised a complex Bayesian multiple hypothesis testing procedure from decision-theoretic principles. Cooper and Sung (1983) also adopted a Bayesian approach using a multiple-window optimal boundary finding algorithm. Recently, Bouthemy (1989) proposed a likelihood ratio hypothesis test for the detection of moving edges. Finally, Kashyap and Eom (1989) also devised a likelihood ratio test for edge-detection in images with more than one texture. This last technique is interesting as it attempts to locate edges by inspection of the data in relatively large segments in adjacent rows/columns of the image. We shall see the relevance of such an approach to our own work in the next section.

We shall consider two aspects of the edge-detection problem; the detection stage itself, and subsequent localization (removal of false edge-points etc). We discuss various aspects of these problems, but will be concerned in particular with a new (Bayesian) approach to the detection stage.

2. Edge-detection in Image Processing

Edge-detection in its broadest sense is a segmentation technique based on the detection of localized discontinuities in an image true scene that arise at texture boundaries. Our interest is in localized discontinuities between larger homogeneous regions, and thus we seek to avoid localized gradient-based techniques that operate over a small grid of pixels. The interpretation of an "edge" at the true scene (unobserved) level is independent of the field-of-vision (entire image or image segment) but this is not the case at the observation level where efficiently learning about an edge usually involves making use of all observations.

2.1 Edge-detection - simple example

Consider the simplest interesting true scene for the edge-detection problem, in which the image field consists of two textures T_1, T_2, labelled by vector parameters θ_1, θ_2 respectively, and separated by a simple edge (defined by a single curve in the plane). In the light of image data **Y**, the goal is to make inferences concerning the location of the simple edge. The inference could take the form of reporting of edge-points in Cartesian coordinates, or of forming some parametric or non-parametric curve to represent the edge, or of presenting a stylised version of the observed image.

As a concrete version of this simple problem, we assume that the image-formation process $f(\theta, \varepsilon)$, corrupts each pixel in the image field independently with additive Gaussian white noise, and that the observed pixel image consists of univariate observations, such that

$$Y_{ij} = \theta_{ij} + \varepsilon_{ij} \qquad \varepsilon_{ij} \sim N(0, \sigma^2). \qquad (4)$$

This is a commonly assumed image-formation model in the analysis of satellite data, and thus the "edge" can be thought of as, for instance, a land-usage boundary, with the image **Y** being the collection of reflectances/radiances in a particular "band" recorded over all pixels. Figure 1a depicts a typical image realization based on the true scene of figure 1b and the above image-formation process. In this 80×80 pixel image, the mean level at every pixel within a texture is equal, with the two different textures having different mean levels ($\theta_1 = 0.0$, $\theta_2 = 1.0$) and common variance ($\sigma^2 = 1.0$).

It is clear from perusal of figure 1a that localized edge-detection techniques that operate over a small sub-grid of pixels do not adequately reflect the nature of the edge-detection problem. A typical edge can only be discerned as such if it marks an abrupt change in some feature of the image between one large region and another. Techniques that do not take this into account cannot, in general, hope to capture the edge structure correctly.

2.2 Changepoint approach to edge-detection

We wish to formulate the edge-detection problem in such a way as to incorporate the notion that an edge should be interpreted as an abrupt (i.e. localized) change in some more large-scale feature. Consider again the simple edge of figure 1b, and a single row (j say) in the image pixel matrix, as depicted in figure 2a. Under the image-formation process (4) and assuming internal homogeneity of textures ($\theta_{ij} = \theta_k$ if pixel (i, j) lies in T_k, $k = 1, 2$) it is clear that the distribution of the observations Y_{ij} in row j is as follows. For some r ($1 \le r \le 80$),

$$Y_{1j}, \ldots, Y_{r_j} \sim N(\theta_1, \sigma^2) \qquad (5)$$

$$Y_{r+1j}, \ldots, Y_{80j} \sim N(\theta_2, \sigma^2)$$

where r represents the unknown (and unobservable) edge-point position in row j. The edge-detection problem now reduces to that of making inference about r in each row. This is the familiar statistical problem of changepoint analysis and identification: see, for example, the reviews of Shaban (1980) and Zacks (1982). Hence the edge-detection problem in image-processing can be reformulated as a practical application of changepoint techniques. Figures 2b and 2c further illustrate this point. Figure 2b is a cross-section of the true level in a single row (row 50) from the true scene of the image in figure 1. It is of the same form as the representations of "ideal edges" in the image-processing literature, with the edge clearly visible between points 40 and 41 on the horizontal scale. Figure 2c is the same row taken from the noise-corrupted image. It is reminiscent of, for instance, time-series plots from system-monitoring operations, an area in which prospective/retrospective identification of changepoints is widely used. We note that the position of the underlying shift in mean-level (i.e. the edge) is barely discernible in figure 2c, due to the noise-corruption, so that localized tests for shift in mean-level would be of little use.

A changepoint approach to edge-detection was in fact loosely suggested by Rosenfeld and Kak (1982), and developed more fully by Basseville (1981). However, Basseville used a prospective scheme derived from Hinkley's cumulative sum procedure (Hinkley (1971)). We feel that a retrospective scheme is more attractive as it reflects the global rather than local aspects of the edge-detection problem.

2.3 Bayesian retrospective changepoint identification

A number of approaches to the changepoint problem appear in the literature, including those based on non-parametric (Pettitt (1980), Hinkley (1971)) or likelihood (Hinkley (1970)) formalisms. The approach we adopt here is Bayesian: see, for example, Chernoff and Zacks (1964), Broemeling (1974), Smith (1975), Booth and Smith (1982), and Stephens (1991). In the Bayesian formulation, inference is made via a posterior distribution for the unknown changepoint position(s), derived from prior assumptions concerning the functional relation between data and population parameters, and prior beliefs about those parameters.

We adopt the following notation. Let $y = (y_1, \ldots, y_n)$ be a realization of a sequence of random variables $Y = (Y_1, \ldots, Y_n)$ of length n. Let θ be the vector of parameters of the sampling distribution, and ψ be a vector of hyperparameters appearing in the specification $[\theta]$ of the prior distribution for θ. The sequence of random variables $Y_1, \ldots, Y_n$ has a changepoint at r ($1 \le r \le n$) if

$$Y_1, \cdots, Y_r \sim [Y_i \mid \theta_1]_1$$
$$Y_{r+1}, \cdots, Y_n \sim [Y_i \mid \theta_2]_2$$

where

$$[. \mid \theta_1]_1 \ne [. \mid \theta_2]_2 .$$

Here, our emphasis will be on retrospective changepoint identification: that is, given $\mathbf{Y}=\mathbf{y}$, our objective is, primarily, to make inferences about the unknown changepoint position, r. Inference will be made via the posterior distribution of r, denoted by $[r \mid \mathbf{Y}, \psi]$. From Bayes theorem, we have

$$[r \mid \mathbf{Y}, \psi] \propto [\mathbf{Y} \mid r, \psi][r]. \tag{6}$$

If θ is wholly or partially unknown, the first term on the right-hand side of (6) is any version of the (integrated) likelihood function obtained from

$$[\mathbf{Y} \mid r, \psi] = \int [\mathbf{Y} \mid r, \theta, \psi][\theta \mid r, \psi],$$

where we adopt the convention that integration is over all variables in the integrand not appearing on the lef-hand side. If θ is completely known the first term on the right-hand side of (6) is simply the likelihood itself (in which case we identify ψ as θ).

By assumption, the conditional distribution of Y given r, θ and ψ is independent of ψ. We shall also assume that $Y_1,...,Y_n$ are conditionally independent given r and θ, and regard θ as independent of r *a priori*. Thus, from (1), the posterior distribution of r is given by

$$[r \mid \mathbf{Y}, \psi] \propto \int \prod_{i=1}^{n} [Y_i \mid r, \theta][\theta][r]. \tag{7}$$

Finally, we often report some estimate of r, such as the posterior mode, $\hat{r}$, say, obtained via $[r \mid \mathbf{Y}, \psi]$, rather than $[r \mid \mathbf{Y}, \psi]$. In the single changepoint case, $[r \mid \mathbf{Y}, \psi]$ is simply a univariate, n-valued discrete distribution, and thus will be easily calculable, with straightforward optimization, moment calculation, etc..

The Bayesian approach extends straightforwardly to the equivalent problems associated with multiple changepoint sequences. We can say that $Y_1,...,Y_n$ has k distinct changepoints at $1 \leq r_1 < r_2 < \cdots < r_k \leq n$ if

$$Y_1, \cdots, Y_{r_1} \sim [Y_i \mid \theta_1]_1$$

$$Y_{r_1+1}, \cdots, Y_{r_2} \sim [Y_i \mid \theta_2]_2$$

$$\cdots$$

$$Y_{r_k+1}, \cdots, Y_n \sim [Y_i \mid \theta_{k+1}]_{k+1}$$

with $\theta=(\theta_1, \cdots, \theta_{k+1})$ having prior distribution $[\theta]$ we make inference via $[r_1, r_2, \cdots, r_k \mid \mathbf{Y}, \psi]$ where

$$[r_1, r_2, \cdots, r_k \mid \mathbf{Y}, \psi] \propto \int \prod_{i=1}^{n} [Y_i \mid r_1, r_2, \cdots, r_k, \theta][\theta][r_1, r_2, \cdots, r_k]. \tag{8}$$

Finally, let M_k denote the model under which we assume precisely k changepoints. In most problems, we will be uncertain about the number of changepoints (if any) that have occurred in any given sequence, but will be able to specify a prior distribution over a finite set of M_ks. In this case, the relevant posterior probabilities are given by (7) multiplied by the prior probability of M_k, with the normalization constant being the summation over all possible k-changepoint models for each M_k. Note that there is an implicit conditioning on M_1 and M_k in (6) and (8) respectively, which we have previously suppressed notationally. Usually, the number of such models considered is small.

In this paper we shall only be concerned with problems where the changepoint(s) arise as the result of parametric rather than distributional changes in the data generation process, that is, where $[\, . \mid . \,]_1 = [\, . \mid . \,]_2$, but $\theta_1 \neq \theta_2$. In any practical problem, implementation reduces to specifying likelihood - prior combinations (corresponding to different image/true scenes), and examining the resulting posterior forms.

2.4 Simple illustration of the edge-detection scheme

We illustrate the implementation of the single changepoint analysis in the context of the simple edge of figure 1a and the image in figure 1b. Consider the elements of row j, say, and let $Y_i \equiv Y_{ij}$. We noted that in (6) the conditional distribution $[Y_i \mid r, \theta]$, $i = 1, \dots, n$, is Normal, with $\theta = (\theta_1, \theta_2, \sigma)$. If we assume that the noise terms are mutually independent for all cells in any row, then $Y_1, \dots, Y_n$ are mutually independent, conditional on θ_1, θ_2, and thus $[Y \mid r, \theta]$ is given by

$$[Y \mid r, \theta] = \prod_{i=1}^{n} [Y_i \mid r, \theta] . \tag{9}$$

For convenience we reparameterize by replacing σ with $\tau = \sigma^{-2}$, so that

$$[Y \mid r, \theta] \propto \tau^{\frac{n}{2}} \exp\left\{ -\frac{\tau}{2} \left[\sum_{i=1}^{r}(Y_i - \theta_1)^2 + \sum_{i=r+1}^{n} (Y_i - \theta_2)^2 \right] \right\} . \tag{10}$$

If the texture mean levels and noise precision are *a priori* unknown we must specify a form for $[\theta]$. For illustration, following Booth and Smith (1982), we choose a simple form of non-informative prior distribution, setting the improper density

$$[\theta] \equiv [\theta_1, \theta_2, \tau] = const \tag{11}$$

for $-\infty < \theta_1, \theta_2 < \infty$, $\tau > 0$. Finally, we specify $[r]$ to be uniform on the range $1 \leq r \leq n-1$. Combining (10) and (11) via (6), we obtain $[r \mid Y, \psi]$ as

$$[r \mid Y, \psi] \propto \left\{ r(n-r) \right\}^{-1/2} \left\{ \sum_{i=1}^{r}(Y_i - \overline{Y_A})^2 + \sum_{i=r+1}^{n} (Y_i - \overline{Y_B})^2 \right\}^{-n/2} \tag{12}$$

where $\overline{Y_A} = \dfrac{1}{r} \sum\limits_{i=1}^{r} Y_i$, and $\overline{Y_B} = \dfrac{1}{n-r} \sum\limits_{i=r+1}^{n} Y_i$.

Note that in this formulation of the edge-detection problem, we consider prior models allowing precisely one changepoint, and thus the valid range for r (under our definition) is $1 \leq r \leq n-1$. Booth and Smith (1982) also consider a "no changepoint" alternative, which extends the valid range for r to $1 \leq r \leq n$, and induces the obvious minor change in (10).

2.5 Edge-detection - results

The posterior density in (12) was evaluated for each row of the image in figure 1a, and the position of the posterior mode recorded, along with the modal probability. The results of this analysis can be seen in figure 3a. It is clear that much of the edge structure has been captured, i.e. many edge-point candidates arising as modes in the changepoint posterior density lie at or close to the true edge-point in the row concerned. In many cases, the results of a preliminary analysis such as this will be sufficiently accurate to enable the subsequent supervised or unsupervised processing techniques to proceed more efficiently. We can easily discern edge-regions as opposed to texture-regions, visually or automatically, allowing for more straightforward segmentation. However, as a representation of the edge itself, figure 3a is obviously inaccurate due to the presence of serious edge misclassifications or "outliers". It is possible to remove these outliers using ideas of spatial continuity of the edge, that is, via our interpretation of the edge itself as continuous in the image field, but for most practical purposes it suffices to consider a simple technique for the removal of such misclassifications.

In the discretized version of the true scene, we would expect edge points in rows and columns to lie close to other edge-points in the adjacent rows and columns. Similarly, we would expect accurate edge-point classifications resulting from an edge-detection analysis to lie in close proximity to each other. Thus any "isolated" candidate points can be regarded as misclassifications, with the term isolated to be defined in some suitable fashion. A possible simple "smoothing" technique (in the sense that isolated candidate points disrupt our interpretation of an edge as being locally continuous at all points on its length) is to centre a small window at each candidate edge-point, and count the number of other edge-points falling within that window. The candidate point can then be accepted as an edge-point or disregarded as a misclassification on the basis of the number of adjacent edge-points. In figure 3b, a 7 x 7 pixel window was used in conjunction with an acceptance criterion of two points per window. As can be seen, a much improved version of the edge results.

For this simple example, and despite the relatively high level of noise-corruption in figure 1a, the changepoint technique for edge-detection has performed reasonably efficiently and effectively. The performance of the technique for lower noise levels (Signal-Noise ratios 1.5, 2.0, 2.5, 3.0) is shown in figures 4a to 4d. The results shown are "unsmoothed".

Retrospective changepoint techniques using data from an entire row seem preferable to localized methods for several reasons. First, as mentioned above, the changepoint approach seems to reflect more adequately the nature of the edge-detection problem. Secondly, the localized methods - differencing, filtering, convolution, local averaging - although intuitively reasonable to some extent are, in fact, generally quite arbitrary. The Bayesian changepoint approach at least has a more secure basis in statistical theory. Thirdly, the localized methods generally depend heavily on expert input of parameters - threshold, window-width etc. - usually arrived at through detailed prior knowledge of the true scene and image. For the changepoint technique, as we have seen, at most only a very general form of prior specification is required. Allied to the last two points, the localized methods return a real number at each cell and rely on thresholding to point up edge-regions, with no measure of uncertainty attached. The changepoint technique returns the most probable edge-position in the row concerned (in light of the data in that row and prior assumptions) together with its associated probability. Finally, and perhaps most importantly, the changepoint technique out-performs the simple localized methods at comparable Signal-Noise ratios, as illustrated in this simple example by figures 4e to 4g, which depict the result of a simple localized edge-detection method. The latter consists of first differencing in two perpendicular directions, with the threshold of acceptance ranging from 5.0 to 3.5. (i.e. we take first-order differences along the rows and columns of the image, evaluate the edge-magnitude at each cell as the square-root of the sum of the squares of these differences, and plot all points for which this magnitude is greater than an arbitrarily chosen threshold value). The Signal-Noise ratio at the edge was 3.0. The results are clearly inferior to those obtained using changepoint analysis on the same image, and are not sufficiently robust to the choice of the (arbitrarily assigned) threshold value.

3. Extensions of the changepoint approach to edge-detection

The above ideas concerning edge-detection via changepoint analysis can be extended in three general directions:

(1) More complex true scenes.

The simple example above demonstrated adequately the use of changepoint techniques in edge-detection. However, although it captured the nature of the edge-detection problem exactly (locating a discontinuity in some aspect of the image arising at the boundary between two non-localised features) it dealt with an over-simple idealised true scene. More realistic true scenes would involve convex objects, multiple regions, patterns, "thin" features etc.. We discuss these in detail below.

(2) Exploitation of spatial continuity.

As observed previously, the analysis of the simple image did not take into account the fact that the edge in the true scene was spatially continuous. It would be reasonable to assume that, in light of the progress made generally in statistical image-processing, the introduction of the notion of local dependence and spatial continuity in the prior would improve results.

(3) Variation of image-formation and noise processes.

In our initial example. we assumed a simple linear form for the image-formation process, and that the noise process corrupted each pixel in the true scene identically and independently with Gaussian white-noise. This again is an idealised situation: observed data might be derived from a different Gaussian, or Poisson, or binary process.

Aspects of these extensions will be explored in the following sections. For a detailed account, see Stephens (1990).

3.1 More complex true scenes

In this section, we seek to extend the formulation and adapt and improve the single changepoint technique so that more complex true scenes may be analysed in a similar fashion.

3.1.1 A simple example

Consider first the class of true scenes where an edge forms a single closed curve, so that the image field is comprised of precisely two texture regions. Clearly, various amendments to our original implementation of the edge-detection scheme are necessary. For illustration, we consider a simple convex object, for example a circle, lying completely within the image field. This test image is inherently different from our original example in that the individual rows and columns contain either zero or two edge-points.

We have seen how to generalize the changepoint formulation from 1 to k changepoints. For convex objects against a homogeneous background, k is at most two, and note further that, in our previous two changepoint notation, the data elements indexed by 1 to r_1 and r_2+1 to n are identically distributed conditional on θ. Assuming the same image-formation and noise processes as in (10), and under the identical prior specification, it can be easily shown that the (exactly) two changepoint equivalent to (12) is given by

$$[r_1, r_2 \mid Y, \psi] \propto ((r_2 - r_1)(n + r_1 - r_2))^{-1/2} \left\{ SSQ_1 + SSQ_2 + SSQ_3 \right\}^{-n/2} \tag{13}$$

where

$$SSQ_1 = \sum_{i=1}^{r_1} (Y_i - \overline{Y_C})^2 \qquad SSQ_2 = \sum_{i=r_1+1}^{r_2} (Y_i - \overline{Y_D})^2 \qquad SSQ_3 = \sum_{i=r_2+1}^{n} (Y_i - \overline{Y_C})^2$$

with

$$\overline{Y_C} = \frac{1}{(n + r_1 - r_2)} \left(\sum_{i=1}^{r_1} Y_i + \sum_{i=r_2+1}^{n} Y_i \right) \quad , \quad \text{and} \quad \overline{Y_D} = \frac{1}{(r_2 - r_1)} \sum_{i=r_1+1}^{r_2} Y_i \ .$$

We can evaluate the posterior distribution (13) for all pairs (r_1, r_2) and locate the joint posterior mode. Zero or one changepoint alternatives correspond to letting r_1 and/or r_2 equal n. For the observed image 5b, derived from 5a, the results of the two changepoint modal posterior analysis is shown in figure 5c.

Note that, however, in the absence of relatively detailed prior knowledge of the true scene, evaluation of the changepoint posterior probabilities for a sequence under the hypothesis of greater than one changepoint is problematic, principally due to the amount of computation involved. For example, rough calculations indicate that the amount of computation required for evaluation of probabilities for the one changepoint model increases linearly with n, whereas the amount of computation for the two changepoint model increases with n^2 (and, similarly, with n^k for the k changepoint model). Clearly, this is prohibitive for large data sequences. The solution to this problem lies in recognizing that the joint changepoint posterior distribution can straightforwardly be computed using the Gibbs sampler.

3.2 Computation via the Gibbs sampler

Suppose that the joint probability structure for a collection of random variables $U_1, ..., U_k$, is such that the joint density $[U_1, ..., U_k]$ is uniquely defined by the full conditional densities $[U_s | U_r, r \neq s]$ for $s = 1, ..., k$. Suppose that samples of U_s can be generated efficiently from $[U_s | U_r, r \neq s]$ given specified values of the conditioning variables, $U_r, r \neq s$. An algorithm for extracting information from these full conditional distributions in order to estimate features of the joint distributions has been discussed by Hastings (1970), and Geman and Geman (1984). This so-called Gibbs sampler algorithm, further developed by Gelfand and Smith (1990), is a Markovian updating scheme which proceeds as follows.

Given arbitrary starting values $U_1^{(0)}, ..., U_k^{(0)}$, we generate a random variate $U_1^{(1)}$ from $[U_1 | U_2^{(0)}, ..., U_k^{(0)}]$, followed by a variate $U_2^{(1)}$ from $[U_2 | U_1^{(1)}, U_3^{(0)}, ..., U_k^{(0)}]$, and so on up to $U_k^{(1)}$ from $[U_k | U_1^{(1)}, ..., U_{k-1}^{(1)}]$. This completes one iteration of the sampling scheme. After t such iterations we would arrive at a joint sample $(U_1^{(t)}, ..., U_k^{(t)})$. As $t \rightarrow \infty$, Geman and Geman show that, under mild regularity conditions, this sample tends in distribution to a variable having joint distribution $[U_1, ..., U_k]$.

Turning specifically to Bayesian applications, suppose that $\phi = (\phi_1, ..., \phi_k)$ is a parameter vector of interest and that given $h(\phi) \propto [\phi | \text{data}]$, we wish to evaluate $[\phi_i | \text{data}]$ for some or all of the $i = 1, ..., k$. Successively regarding $h(\phi)$ as a function of ϕ_i for fixed ϕ_j, $j \neq i$, immediately identifies functions $h_i(\phi_i | \phi_j, j \neq i)$ for $i = 1, ..., k$, which are such that

$$h_i(\phi_i | \phi_j, j \neq i) \propto [\phi_i | \phi_j, j \neq i, \text{data}].$$

From knowledge of $h_i(\phi_i | \phi_j, j \neq i)$, we can sample from $[\phi_i | \phi_j, j \neq i, \text{data}]$, so that the Gibbs sampling algorithm is seen to provide a general solution to the problem of summarizing the posterior distribution the specification of likelihood and prior.

Now, consider the sequence of random variables $Y = (Y_1, \ldots, Y_n)$ assumed to have k changepoints $(r_1, \ldots, r_k)$ of unknown position but where k is presumed known. Let $r_0 = 0$, and $r_{k+1} = n$. Let the sampling distribution of $Y_j = (Y_{r_{j-1}+1}, \ldots, Y_{r_j})$ have parameters θ_j, and $(Y_{r_{j-1}+1}, \ldots, Y_{r_j})$ be conditionally independent given θ_j, $j = 1, \ldots, k+1$. Finally, let $(Y_1, \ldots, Y_{k+1})$ be conditionally independent given $\theta = (\theta_1, \ldots, \theta_{k+1})$.

In order to implement the Gibbs sampler, we must have explicit forms (up to proportionality) for the set of full conditional posterior distributions for the unknown parameters. From the conditional independence assumptions, we may write the marginal posterior distribution of r_j conditional $(r_0, \ldots, r_{j-1}, r_{j+1}, \ldots, r_{k+1})$, as

$$[r_j \mid r_0, \ldots, r_{j-1}, r_{j+1}, \ldots, r_{k+1}, Y, \psi] \equiv [r_j \mid r_{j-1}, r_{j+1}, Y_j, Y_{j+1}, \psi]$$

where

$$[r_j \mid r_{j-1}, r_{j+1}, Y_j, Y_{j+1}, \psi] \propto \int \prod_{i=r_{j-1}+1}^{r_j} [Y_i \mid \theta_j] \prod_{i=r_j+1}^{r_{j+1}} [Y_i \mid \theta_{j+1}] [\theta_j, \theta_{j+1} \mid \psi] . [r_j \mid r_{j-1}, r_{j+1}] . \tag{14}$$

It is clear that in this formulation of the multiple changepoint problem, which is identical to our original one, the marginal posterior distribution for r_j conditional on the other $k-1$ changepoints depends only on r_{j-1} and r_{j+1}, for $j = 1, \ldots, k$. But conditional on r_{j-1} and r_{j+1}, the marginal posterior distribution for r_j is identical to the usual one changepoint posterior distribution given by (12) evaluated for the sub-sequence (Y_j, Y_{j+1}), with the valid range of r_j being restricted to $r_{j-1}+1, \ldots, r_{j+1}$. Thus we can identify $[r_j \mid r_{j-1}, r_{j+1}, Y, \psi]$ using standard and familiar techniques, for $j = 1, \ldots, k$.

Note that in this formulation we have integrated out the parameters of secondary interest, namely $\theta = (\theta_1, \ldots, \theta_{k+1})$. If the θ_j's were of interest, and we wanted to calculate the marginal posterior densities $[\theta_j \mid Y, \psi]$, we could extend the Gibbs sampler by including the $k+1$ conditional posterior densities

$$[\theta_j \mid r_0, \ldots, r_{k+1}, \theta_1, \cdots, \theta_{j-1}, \theta_{j+1}, \cdots, \theta_{k+1}, Y, \psi]$$

in the sampling cycle described above. It is easily seen that this conditional posterior density for θ_j simplifies to

$$[\theta_j \mid r_{j-1}, r_j, Y_j, \psi] \propto \int \prod_{i=r_{j-1}+1}^{r_j} [Y_i \mid \theta_j][\theta_j \mid \psi] \tag{15}$$

We would now have now simulate $2k+1$ observations per iteration, as opposed to k previously, so we might expect processing time to increase, at least by a factor of two, and by a greater factor if the functional form of $[\theta_j \mid r_{j-1}, r_j, Y_j, \psi]$ is difficult to sample from. Also convergence may be more difficult to discern for the marginal posterior densities for the "continuous" parameters θ_j.

Thus, we might expect the amount of computation per iteration of the Gibbs sampler for the k changepoint problem to be of order nk, rather than the order n^k of the exact technique. Also, hopefully, only a small number of iterations will be required to obtain a reasonable approximation to the joint posterior density, and even fewer needed to locate the position of the joint posterior mode. These points have been confirmed in the case of the normal data-generation model for various k and large data sequences by means of an extensive experimental study. The changepoint-based edge-detection technique is therefore computationally feasible even for large complex images.

3.3 Spatial dependence and edge continuity

The (prior) model for the unobservable true scene on which we have based our simple analysis fails to reflect the important prior knowledge that regions are spatially contiguous, but "textured". Other image processing techniques usually encapsulate this prior opinion by means of a Markov Random Field prior distribution specified locally for pixels in the true scene. This, in fact, is similar to the model we have used, but with another stage in the model hierarchy. To perform changepoint-based edge-detection with such a prior model is theoretically possible, but in practice often becomes computationally prohibitive. Again, however, it would appear that a Gibbs sampler approach is of use, as, conditionally, the changepoint full conditionals involved are only slightly more complex than those derived under the simpler model. In the edge-detection context, we also have prior knowledge relating to edge continuity. For example, we know that, in general, edges are continuous, and that edge-points in adjacent rows will lie close together. This prior knowledge can be easily introduced in a sequential analysis via the prior distribution for changepoint position (see Stephens (1990)).

3.4 Variation of image-formation process

Our discussion thus far has focussed on the case of a linear model with additive noise. More generally, we might have

$$Y_{ij} = \theta_{ij} * \varepsilon_{ij}$$

where $*$ is some operation, such as addition or multiplication etc., and where the ε_{ij} are not identically distributed for all i,j. In such cases, provided we can still embody the image-formation process in terms of functional forms for likelihoods, the changepoint-based technique can be applied in precisely the same way as before. For example, consider an additive image-formation model of the form

$$Y_{ij} = \theta_{ij} +_2 \varepsilon_{ij} \tag{16}$$

with θ_{ij} taking the values 0 or 1, $+_2$ representing addition modulo 2, and with ε_{ij} taking the values 0 or 1 with probabilities $1-p$ and p respectively, p being an *a priori* unknown parameter in the model. This describes precisely the additive binary noise model. If we again assume that the error terms are independent, the observed values Y_{ij} are conditionally independent given the true scene values θ_{ij}. Now, consider as before a single row, j say, taken from the image data. It can be seen that the conditional distribution of data elements in row j is given by

$$[Y_{ij} \mid r, \theta_{ij}, p] \equiv \begin{cases} Bernoulli\,(p) & i = 1, \ldots, r \\ Bernoulli\,(1-p) & i = r+1, \ldots, n \end{cases} \tag{17}$$

where if we allow p to take values on the whole of $(0,1)$, (17) reflects our *a priori* indifference as to whether texture 1 "precedes" texture 2 in the underlying true scene (of a form like that of Figure 1b). Suppressing the dependence on j, it is clear that the likelihood $[\mathbf{Y} \mid r, p]$ is given by

$$[\mathbf{Y} \mid r, p] = p^{n-r+S_r}(1-p)^{r-S_r} \tag{18}$$

where $S_r = \sum_{i=1}^{r} Y_i - \sum_{i=r+1}^{n} Y_i$. Assuming a uniform prior for r over $1, \ldots, n-1$, and a prior for p of the form

$$[p \mid \psi] \propto (p(1-p))^{-\frac{1}{2}},$$

the posterior distribution of r is given by

$$[r \mid \mathbf{Y}, \psi] \propto \Gamma(n-r+S_r+0.5)\,\Gamma(r-S_r+0.5) \tag{19}$$

The changepoint technique can also be extended to incorporate data originating from Poisson sources. Here, however, it might be advantageous to make a variance-stabilizing square-root transformation of the data to normality.

3.5 Object detection

The emphasis in object detection is somewhat different to that of general edge-detection since we might have specific knowledge that one or more objects of small diameter (in terms of numbers of pixels) are present in a much larger, homogeneous background.

To take a simple example, suppose a single, small object of maximum pixel diameter R and unknown orientation is present in a true scene, and that the object in the image has raised mean signal level θ_2 on a background having mean signal level θ_1, with noise-variance σ^2. Then, in the multiple changepoint framework, again taking each row/column in the image matrix individually, this is simply a two changepoint detection problem, except that the distance between the changepoints is known to be less than or equal to the maximum dimension of the object, so that $r_2 - r_1 \leq R$. Thus we may use a restricted version of (13) to detect the object, computing the joint posterior probability for all pairs of r_1 and r_2 that satisfy this constraint, and recording the modal value. Such an operation may, however, be quite computationally demanding if R is "too large", and an excessive number of pairs of changepoints have to be considered. For most practical purposes it suffices to consider only values precisely $m < R$ apart, for some suitable m.

4. Illustrations

We now present examples of the changepoint methodology described above as applied to various simulated and real data images.

Figure 6a is an image derived from the artificial multi-region true scene in figure 6b. Each separate template has a level 2.0 against the background of level 0.0, with unit normal white-noise superimposed to form the image. Figure 6c shows the (smoothed) results of a Gibbs sampler-based four changepoint analysis of the image, with posterior modal positions for each changepoint recorded. The darker the symbol plotted at each location, the higher its (estimated) posterior probability. The analysis of the 80×80 image took around half a minute on a SUN SPARCstation.

Figure 7a is 64×64 two-level representation of Ireland taken from Ripley (1987). Figure 7b is the derived image, with assumed mean levels 0.0 and 1.0 and noise of standard deviation 0.65 superimposed. Figure 7c shows the results of a two changepoint analysis, which took 20 seconds to perform.

Figure 8a is a 256×256 real data facial image. Figure 8b shows the results of a five changepoint analysis of the image assuming a (probably inappropriate) normal noise process. The outline of hair, face, and shoulders have been detected adequately. Other smaller features (shadowing, neckline, facial features) have been incompletely delineated, but this is partly to be expected, due to the assumptions on which the changepoint model is based. The analysis took three minutes.

Figure 9a is an artificial multi-object image, where the objects are of level 2.0 on a background of level 0.0. Figure 9b shows the results of an exact analysis using the restricted two changepoint model as descibed in section 3.5. The results are smoothed in accordance with our prior knowledge that the changepoints will occur more densely locally than in the edge-detection context. The analysis and smoothing took 1.5 seconds to perform.

5. Discussion

In this paper, we have described the Bayesian approach to retrospective changepoint identification applied to edge-detection, formulated a Gibbs sampler-based computational approach, and showed that this algorithm could produce results to an acceptable degree of accuracy in a reasonable amount of computational time.

The major reason for adopting a Gibbs sampler-based approach in the solution of any statistical problem is computational convenience and/or necessity. Changepoint problems are especially suitable for solution via the Gibbs sampler due to the simple conditional structures involved, and, as we have seen in the multiple changepoint case, use of the Gibbs sampler can often result in considerable computational savings.

Acknowledgements

Part of this research was carried out with support from the UK Science and Engineering Research Council's Complex Stochastic Systems Initiative. We thank British Telecommunications Research Laboratories, Martlesham Heath, for the data-base from which Figure 8a is derived. A version of this paper was presented at the Twelfth Rencontre Franco-Belge de Statisticiens held at Louvain-la-Neuve, Belgium, 21-22 November, 1991.

References

Basseville, M. (1981), Edge detection using sequential methods for change in level - Part II: Sequential detection of change in mean, *IEEE Transactions on Acoustics, Sound, and Signal Processing*, **ASSP-29**, 32-50.

Booth, N. B., and Smith, A. F. M. (1982), A Bayesian Approach to Retrospective Identification of Change-points, *Journal of Econometrics*, **19**, 7-22.

Bouthemy, P. (1989), A maximum likelihood framework for determining moving edges, *IEEE Transactions on Pattern Analysis and Machine Intelligence*, **PAMI-11**, 499=-511

Broemeling, L. D. (1974), Bayesian Inferences about a Changing Sequence of Random Variables, *Communications in Statistics, Part A - Theory and Methods*, **3**, 243-255.

Canny, J. (1986), A computational approach to edge detection, *IEEE Transactions on Pattern Analysis and Machine Intelligence*, **PAMI-8**, 679-698

Chernoff, H., and Zacks, S. (1964), Estimating the Correct Mean of a Normal distribution which is subjected to a Change in Time, *Annals of Mathematical Statistics*, **35**, 999-1018.

Cooper, D. B., and Sung, F. P. (1984), Multiple-window parallel adaptive boundary finding in computer vision, *IEEE Transactions on Pattern Analysis and Machine Intelligence*, **PAMI-5**, 299-314

Gelfand, A. E., and Smith, A. F. M. (1990), Sampling Based Approaches to Calculating Marginal Densities, *Journal of the American Statistical Association*, **85**, 398-409.

Geman, S., and Geman, D. (1984), Stochastic relaxation, Gibbs distributions and the Bayesian restoration of images, *IEEE Transactions on Pattern Analysis and Machine Intelligence*, **PAMI-6**, 721-741

Hinkley, D. V. (1970), Inference about the Changepoint in a Sequence of Random Variables, *Biometrika*, **57**, 1-17

Hinkley, D. V. (1971), Inference about the Change-point from Cumulative Sum Tests, *Biometrika*, **58**, 509-523

Kashyap, R. L., and Eom, K.-B. (1989), Texture boundary detection based on the long correlation model, *IEEE Transactions on Pattern Analysis and Machine Intelligence*, **PAMI-11**, 58-67

Lu, Y., and Jain, R. C. (1989), Behavior of edges in scale space, *IEEE Transactions on Pattern Analysis and Machine Intelligence*, **PAMI-11**, 337-356

Mascarenhas, N. D. A., and Prado, L. O. C. (1980), A Bayesian approach to edge detection in images, *IEEE Transactions on Automatic Control*, **AC-25**, 36-43

Pettitt, A. N. (1980), A simple Cumulative Sum type Statistic for the Changepoint Problem with Zero-One observations, *Biometrika*, **67**, 79-84

Ripley, B. D. (1987), *Stochastic Simulation*, Wiley, New York.

Rosenfeld, A., and Kak, A. C. (1982), *Digital Picture Processing*, Academic Press, New York.

Shaban, S. A. (1980), Change point problem and two-phase regression: an annotated bibliography, *International Statistical Review*, **48**, 83-93

Smith, A. F. M. (1975), A Bayesian Approach to Inference about a Changepoint in a Sequence of Random Variables, *Biometrika*, **62**, 407-416

Stephens, D. A. (1990), *Bayesian Edge Detection in Image Processing*, Unpublished Ph.D. thesis, University of Nottingham.

Stephens, D. A. (1991), Bayesian Retrospective Multiple Changepoint Identification, Submitted for publication to Applied Statistics.

Torre, V., and Poggio, T. A. (1986), On edge detection, *IEEE Transactions on Pattern Analysis and Machine Intelligence*, **PAMI-8**, 147-163

Zacks, S. (1982), Classical and Bayesian approaches to the changepoint problem, *Statistique et Analyse des Donnees*, **1**, 48-81

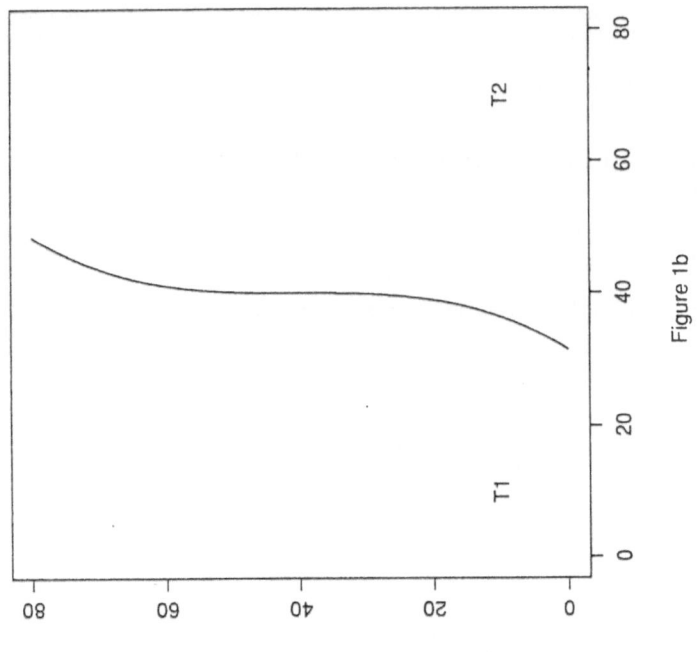

Figure 1b

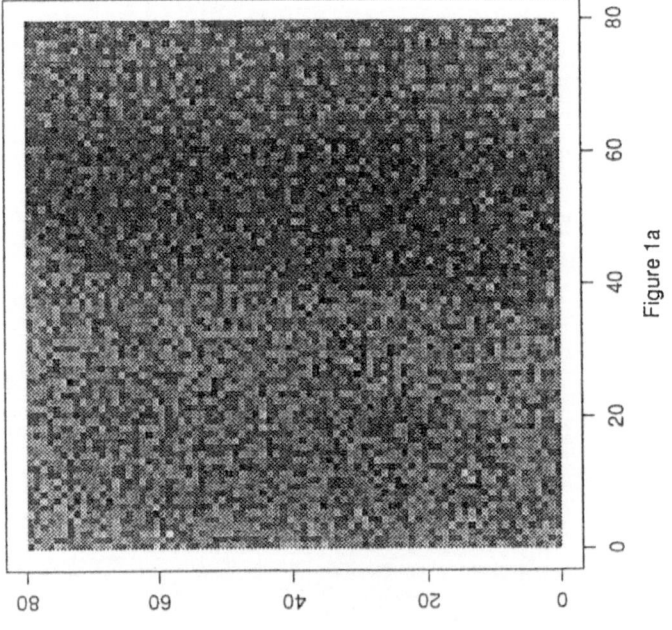

Figure 1a

21

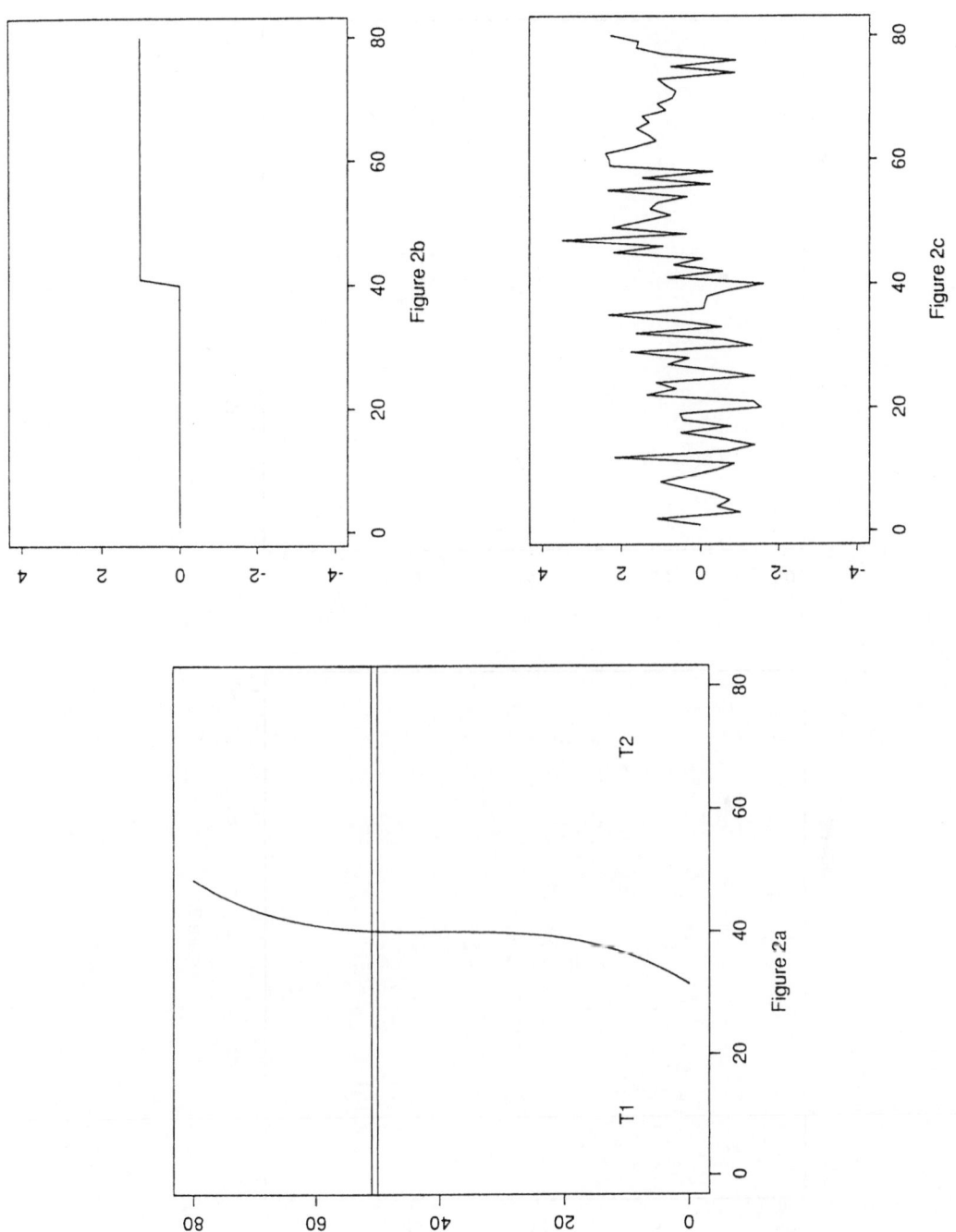

Figure 2b

Figure 2c

Figure 2a

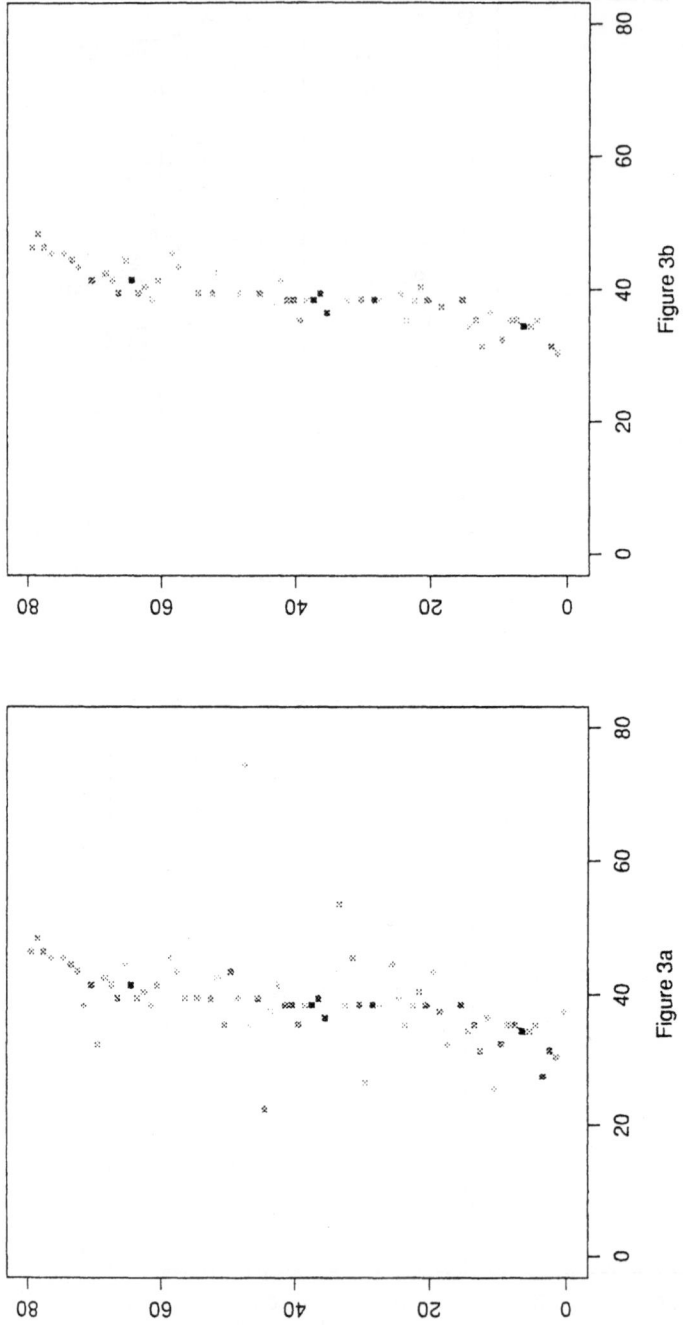

Figure 3a

Figure 3b

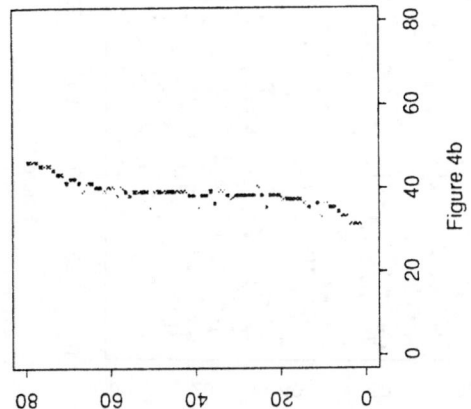

Figure 4b

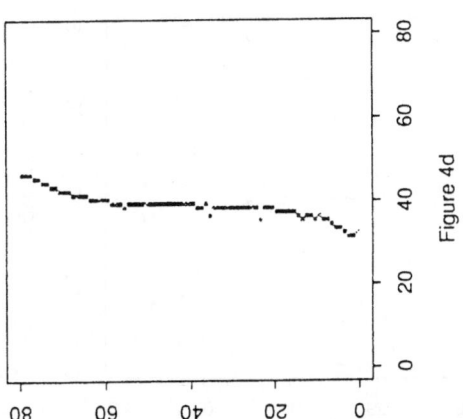

Figure 4d

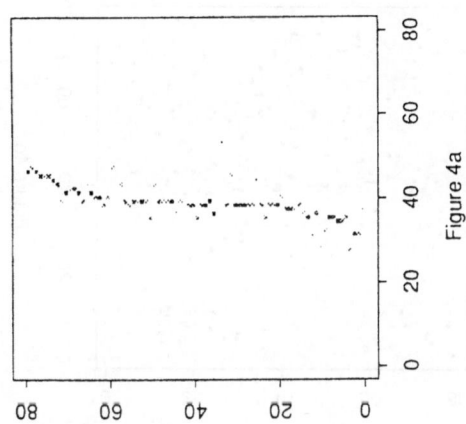

Figure 4a

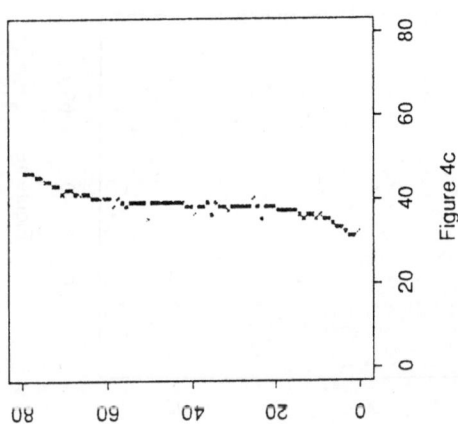

Figure 4c

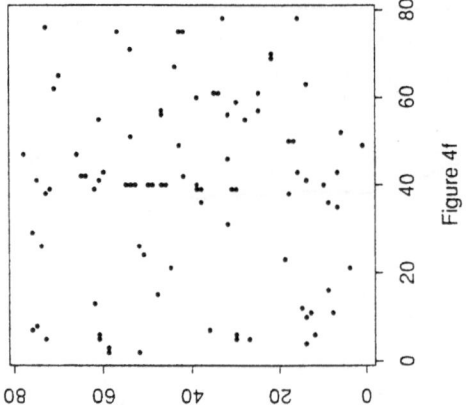

Figure 4f

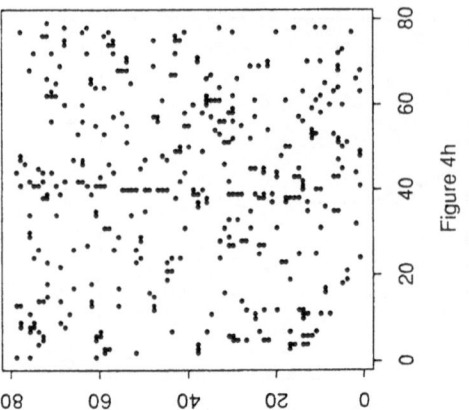

Figure 4h

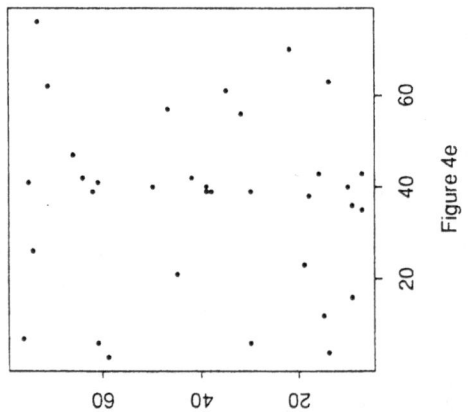

Figure 4e

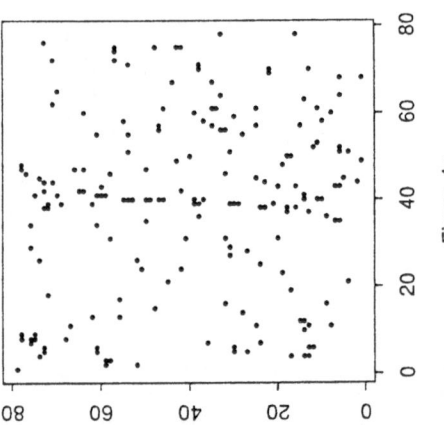

Figure 4g

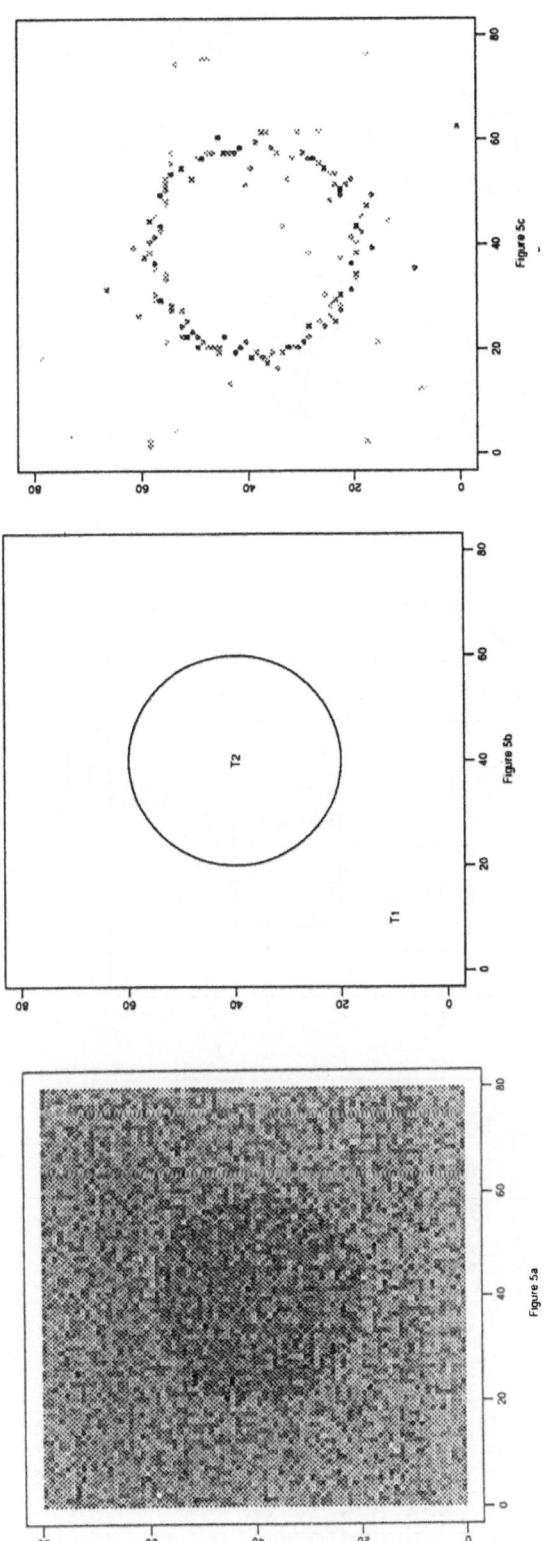

Figure 5a

Figure 5b

Figure 5c

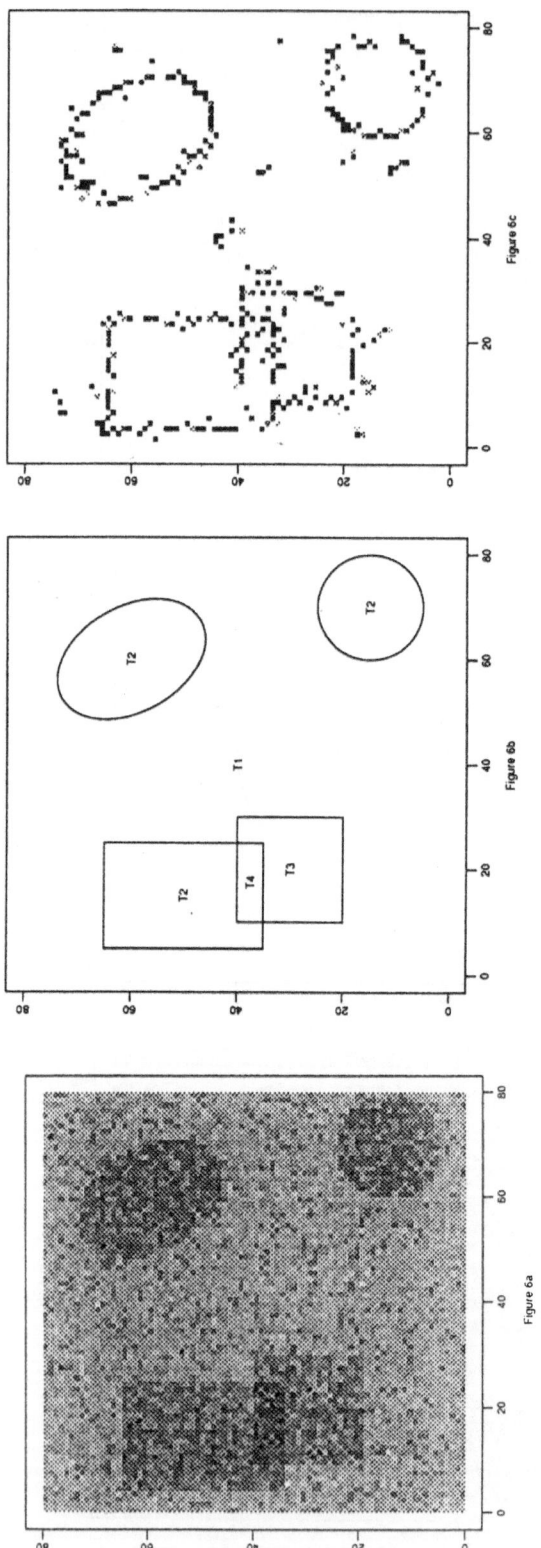

Figure 6a

Figure 6b

Figure 6c

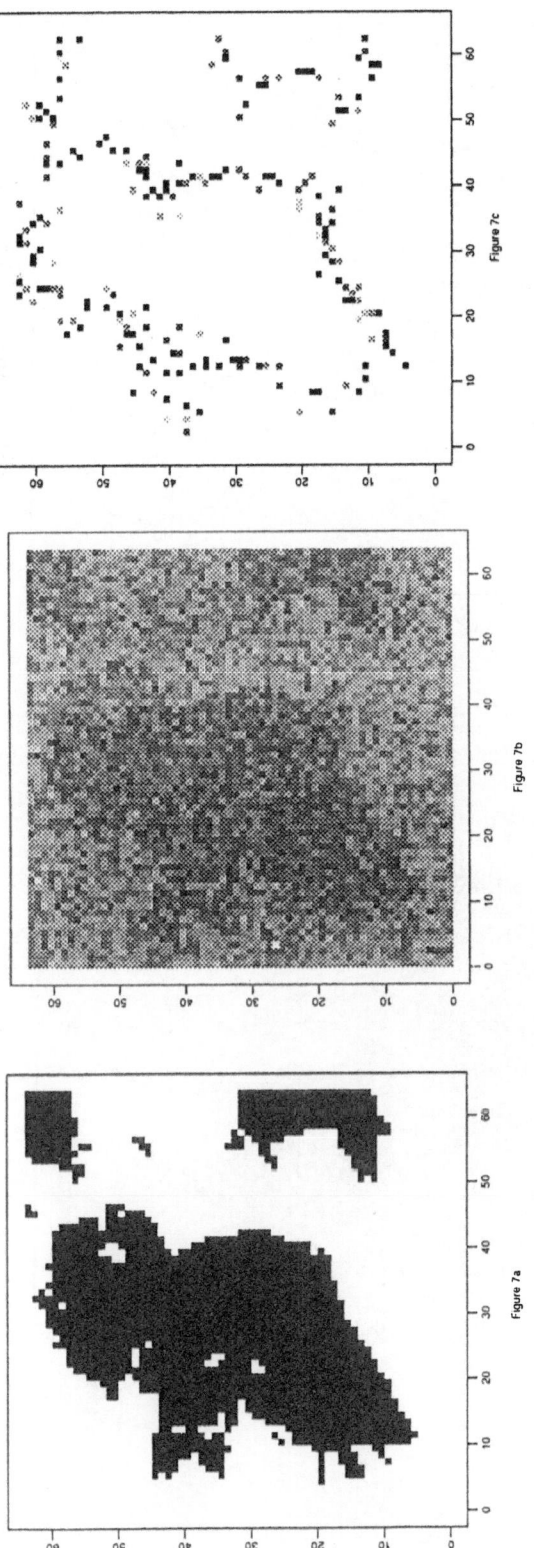

Figure 7c

Figure 7b

Figure 7a

28

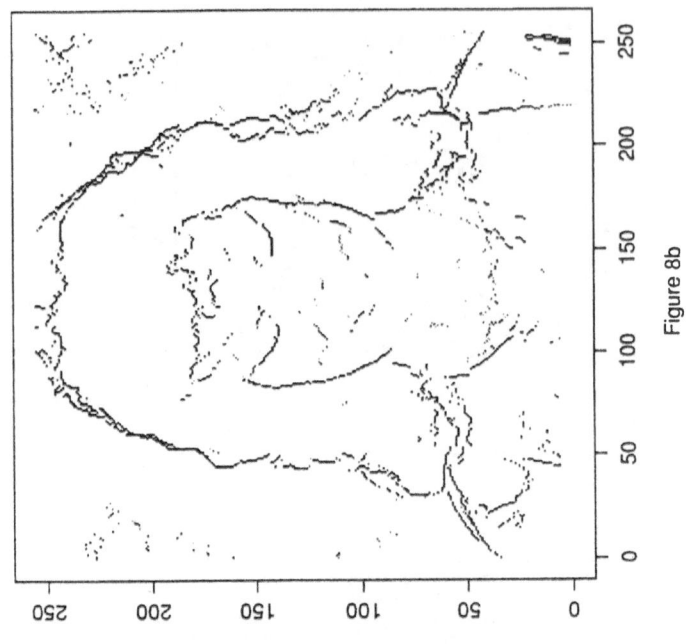

Figure 8b

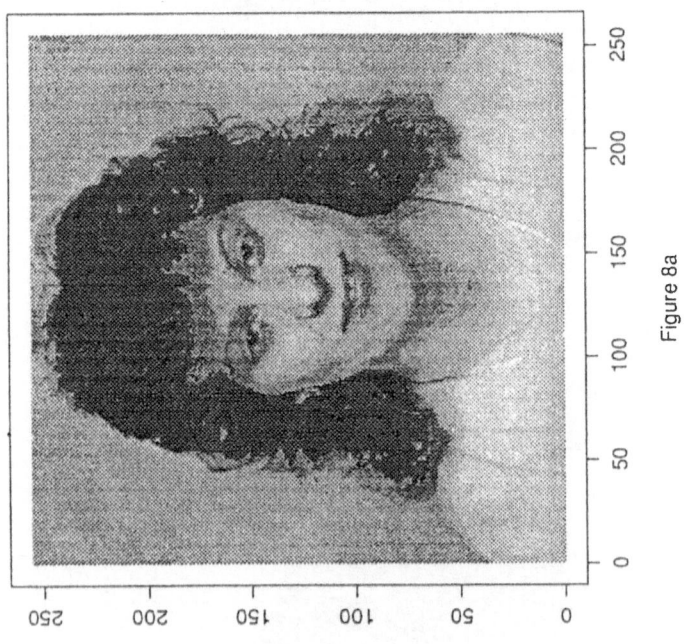

Figure 8a

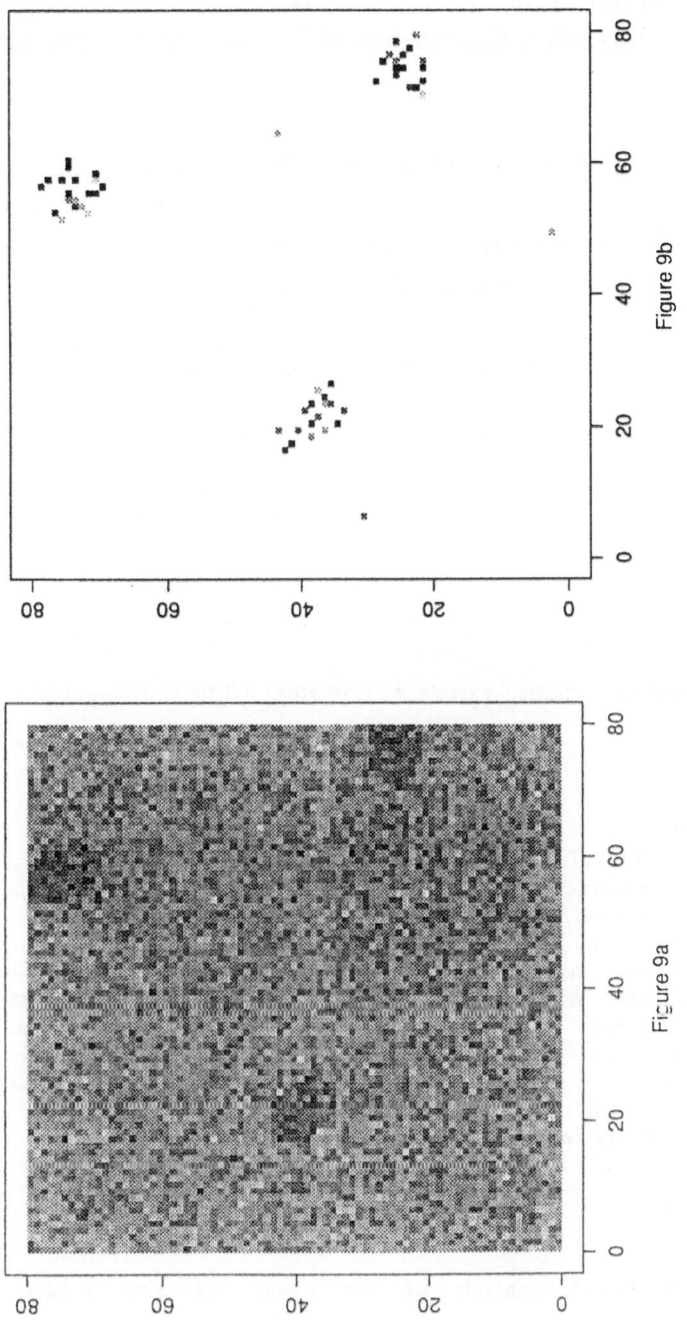

Figure 9a

Figure 9b

Efficient Computer Generation of Matric-Variate *t* Drawings with an Application to Bayesian Estimation of Simple Market Models

Frank Kleibergen and Herman K. van Dijk

Econometric Institute and Tinbergen Institute
Erasmus University Rotterdam
P.O. Box 1738
3000 *DR Rotterdam, The Netherlands*

Keywords : matric–variate *t*, (inverted) Wishart, triangularisation,
Simultaneous Equations Model.

Abstract

Algorithms for efficient computer generation of matric-variate *t* random drawings are constructed which make use of two results in distribution theory. First, the definition of a matric-variate *t* distributed random matrix as the product of a matric-variate normal distributed random matrix and the square root of an inverted-Wishart distributed random matrix. Second, a decomposition of the Wishart and inverted Wishart matrix into triangular matrices. The different steps of the algorithm for matric-variate *t* drawings and the decomposition of the (inverted-) Wishart are explained. For illustrative purposes, the posterior density of the structural parameters of a simple market model is evaluated. These structural parameters are nonlinear functions of matric-variate *t* variables.

1. Introduction

Bayesian statistical analysis of the linear regression model, denoted by $y = X\beta + \varepsilon$, leads to the wellknown result that, given a uniform prior, the marginal posterior density of the parameters of interest β is a *multivariate* *t* density; see, *e.g.*, Raiffa and Schlaiffer (1961, Chapter 13), Zellner (1971, Ch. 3), Box and Tiao (1973, Ch. 2.7) or Judge et al (1985, Ch. 4). Several properties of this density cannot be analyzed by standard analytical integration methods. The use of Monte Carlo numerical integration is becoming a standard technique in this context. For instance, using Monte Carlo integration, Geweke (1986) evaluates the effect of linear inequality restrictions on the posterior density of β. Further, Van Dijk and Kloek (1980) and DeJong and Whiteman (1991) use Monte Carlo integration to study the distribution of a nonlinear function of the parameter vector β. Specifically, within the context of an autoregressive model these authors compute the density of the roots of the characteristic polynomial. These roots are nonlinear functions of the parameter vector β.

In the multivariate regression model, usually denoted by $Y = X\Pi + V$, one deals with a matrix Y of random variables and with a matrix of coefficients Π. Within this class of models, it is wellknown that, given a uniform prior, the marginal posterior density of Π is a *matric–variate* t density; see, *e.g.*, Zellner (1971, Ch. 8), Box and Tiao (1973, Ch. 8). Nonlinear functions of the parameters Π are analyzed using Monte Carlo integration methods by Geweke (1988) and by Zellner, Bauwens and Van Dijk (1988) and can be computed using the computer packages RATS (Ch. 10) and SISAM (Hop and Van Dijk (1992)). In the present paper we give a detailed explanation of the different steps of an efficient algorithm for the generation of *matric–variate* t drawings. In particular, we focus on a simple sequence of steps that lead to a decomposition of a positive definite symmetric matrix of *Wishart* and *inverted–Wishart* random variables into triangular matrices. For illustrative purposes, we make use of two simple market models where the parameters of interest are nonlinear functions of *matric–variate* t variables.

The contents of this paper are organized as follows. In the next section we discuss the different steps that lead to a *matric–variate* t distribution. In section 3 a decomposition of the *Wishart* and *inverted–Wishart* density is explained in detail and a flow diagram summarizes the generation of a matrix of random drawings that is defined as a lower triangular square root of a (inverted–) Wishart distributed random matrix. In section 4 the illustration using market models is discussed. The results of this paper show that classical asymptotic results may be very inaccurate as approximation of the exact finite sample results of a Bayesian analysis.

2. Relations between the matric-variate normal, (inverted-) Wishart, and matric-variate t distributions

Two well-known functions of standard normal distributed random variables are :

1. the χ^2 random variable with λ degrees of freedom, defined as the sum of squares of λ independent normal distributed random variables;

2. the t random variable with λ degrees of freedom, defined as the ratio of a normal and the square root of an independent χ^2 random variable with λ degrees of freedom; see, *e.g.*, Hogg and Craig (1978), or DeGroot (1987).

These relationships between univariate random variables can be extended to the case of random matrices. In figure 1 the relations between the matric-variate normal, (inverted-) Wishart and matric-variate t distributions are sketched together with the transformation functions of the random matrices. For convenience, we consider the univariate case. Let x and z be standard normal distributed, *i.e.*, $x \sim N(0, 1)$ and $z \sim N(0, 1)$. It follows from a simple transformation of variables that $y = x^2$ has a χ^2 distribution with one degree of freedom and that $v = y^{-1}$ is inverted χ^2. The standard t random variable with one degree of freedom (Cauchy) is defined through the transformation function $t = zv^{\frac{1}{2}}$ and the t density is defined as the marginal density of an uncountable mixture of conditional normal densities with an inverted χ^2 density as mixing density. For details we refer to textbooks in statistics as Raiffa and Schlaiffer (1961, pp. 232-233).

A nonstandard t distribution may be defined through a linear transformation function which changes the location and scale of the random variable. For instance, let $t \sim t(0, 1, \lambda)$ then $t^* = \sigma t + \mu$ with t^* again a t distributed random variable with location parameter μ and scale parameter σ.

It follows directly from the relations presented in Figure 1 that, given a standard normal drawing and an inverted χ^2 drawing, the generation of a t drawing is trivial. Efficient techniques for this are discussed in Kinderman and Monahan (1980). The extension of these results from the univariate case to the multivariate case is straightforward. One takes a vector z of standard normal drawings. The other steps remain the same and one obtains a vector of t random drawings. Here the transformation to a nonstandard distribution involves a scale transformation with the square root of a positive definite scale matrix. For details, see Zellner (1971, App. A) and Anderson (1984).

For the matric-variate case one has the extension of the univariate standard normal of x and z to matrices X and Z of standard normals. Further, one needs the distribution of the inner product $Y = X'X$ of a matrix of standard normal variables X and the inverse of Y, defined as $V = Y^{-1}$. The derivation of these results is given in, *e.g.*, Bauwens (1984) and Press (1972). For the computer generation of matric-variate t random variables one needs only a

3. Algorithms for generating a triangular square root of Wishart and inverted-Wishart distributed random matrices

We start with some results for the standard matric-variate normal distribution. Let X be a $\lambda \times m$ matrix, partitioned as

$$X = (X_1\ X_2), \text{ with } X_1\text{: } \lambda \times n, \text{ and } X_2\text{: } \lambda \times l,\ l + n = m$$

The matrix X has a standard matric-variate normal probability density function (pdf) denoted by

$$p(X) = f_{Mn}(X|\ 0,\ I_m \otimes I_\lambda)$$

For details we refer to Press (1972). Since the scale matrices are identity matrices, the decomposition of the matrix normal distribution of X into a conditional normal of X_2, given a value of X_1, and a marginal distribution of X_1 is straightforward. Uncorrelatedness implies independence in the case of the normal distribution. Therefore we have $p(X) = p(X_1)p(X_2)$ and $p(X_2|X_1) = p(X_2)$, where

$$p(X_1) = f_{Mn}(X_1|\ 0,\ I_n \otimes I_\lambda) \text{ and } p(X_2|X_1) = f_{Mn}(X_2|\ 0,\ I_l \otimes I_\lambda)$$

A random matrix Y with a Wishart distribution with λ degrees of freedom and identity scale matrix can be defined as the innerproduct of the matric-variate standard normal distributed $\lambda \times m$ random matrix X, defined above, see Anderson (1984) and Press (1972). Then one has

$$Y = X'X \text{ with } p(Y) = f_W(Y|\ I_m,\ \lambda),\ \lambda > m - 1$$

As a next step we consider the reduction of the square matrix Y into an upper triangular matrix by partitioning Y and by postmultiplying Y by a lower triangular matrix

$$Y = \begin{bmatrix} Y_{11} & Y_{12} \\ Y_{21} & Y_{22} \end{bmatrix},\ Y\begin{bmatrix} I_n & 0 \\ -Y_{22}^{-1}Y_{21} & I_l \end{bmatrix} = \begin{bmatrix} Y_{11.2} & Y_{12} \\ 0 & Y_{22} \end{bmatrix},\ Y_{11.2} = Y_{11} - Y_{12}Y_{22}^{-1}Y_{21}$$

The determinant of the lower triangular matrix equals 1, which implies that the jacobian of the transformation from $(Y_{11},\ Y_{12},\ Y_{22})$ to $(Y_{11.2},\ Y_{12},\ Y_{22})$ also equals 1. Given the symmetry of Y $(Y_{12} = Y'_{21})$ the probability density function of the upper triangular matrix equals the pdf of the original matrix as a consequence.

The marginal and conditional pdf's of $(Y_{11.2},\ Y_{12},\ Y_{22})$ can be constructed using the decomposition of the matric-variate normal pdf. Consider

$$p(X) = f_{Mn}(X|\ 0,\ I_m \otimes I_\lambda),\ Y = X'X \Rightarrow p(Y) = f_W(Y|\ I_m,\ \lambda),\ \lambda > m-1$$

matrix of standard normal drawings and the so-called square root of the inverted-Wishart matrix of random drawings. An efficient technique to perform this is discussed in the next section.

Figure 1. Relations between matrix normal, (inverted-) Wishart, and matrix-t distributions.

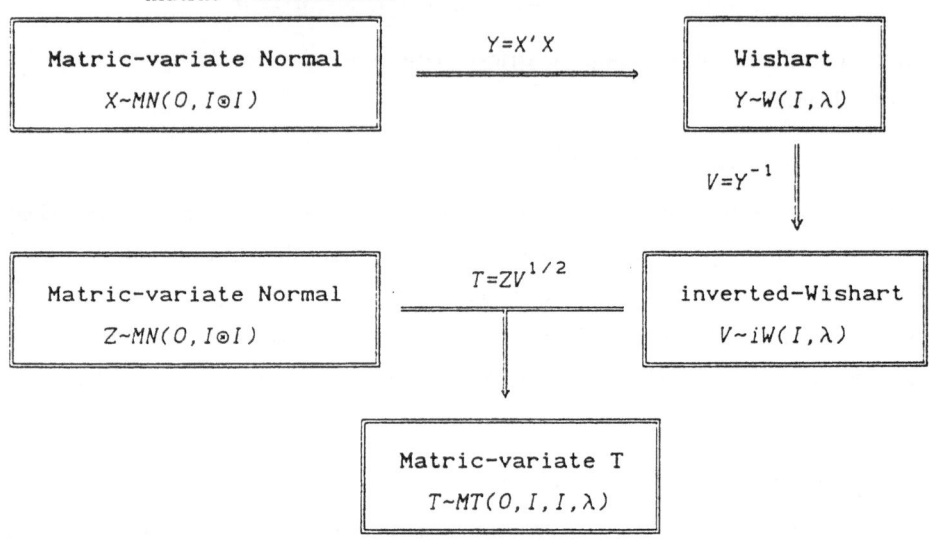

X : $\lambda \times m$ matrix

Y : $m \times m$ matrix

Z : $k \times m$ matrix

The transformation of a standard matric-variate t to a nonstandard t through a location-scale transformation is as follows. Let T be a matrix of standard matric-variate t drawings, then one can use

$$T^* = ATB + \mu$$

as transformation function. As a result T^* has a matric-variate T distribution with parameters μ, $A'A$, $B'B$, and the same degrees of freedom as before when A and B are both full rank square matrices; see, *e.g.*, Bauwens (1984), Drèze and Richard (1983).

$$X = (X_1 \ X_2), \quad X_1 : \lambda \times n, \quad X_2 : \lambda \times l, \quad m = n+l, \quad M_{X_2} = I_\lambda - X_2(X_2'X_2)^{-1}X_2'$$

The triangular decomposition defined above results in

$$Y = X'X \Rightarrow \begin{bmatrix} Y_{11} & Y_{12} \\ Y_{21} & Y_{22} \end{bmatrix} = \begin{bmatrix} X_1'X_1 & X_1'X_2 \\ X_2'X_1 & X_2'X_2 \end{bmatrix} \Rightarrow \begin{bmatrix} Y_{11.2} & Y_{12} \\ 0 & Y_{22} \end{bmatrix} = \begin{bmatrix} X_1'M_{X_2}X_1 & X_1'X_2 \\ 0 & X_2'X_2 \end{bmatrix}$$

Given the marginal and conditional distributions of X_1 and X_2 it is possible to construct the marginal and conditional distributions of (i) Y_{22}, (ii) Y_{21} ($= Y_{12}'$) and (iii) $Y_{11.2}$.

(i) Given $p(X_2) = f_{Mn}(X_2| \ 0, \ I_l \otimes I_\lambda)$ and $Y_{22} = X_2'X_2$, application of the definition of a Wishart distributed random matrix yields

$$p(Y_{22}) = p(X_2'X_2) = f_W(Y_{22}| \ I_l, \ \lambda) \Rightarrow Y_{22} \sim W(I_l, \lambda)$$

(ii) Given $p(X_1) = f_{Mn}(X_1| \ 0, \ I_n \otimes I_\lambda)$, $Y_{21} = X_2'X_1$ and the linear combination property of the matric-variate normal, it follows that the conditional pdf of $Y_{21} = X_2'X_1$ given a value of X_2 equals

$$p(X_2'X_1|X_2) = f_{Mn}(X_2'X_1| \ 0, \ I_n \otimes X_2'X_2) \Rightarrow p(Y_{21}|Y_{22}) = f_{Mn}(Y_{21}| \ 0, \ I_l \otimes Y_{22})$$

$$Y_{21}|Y_{22} \sim MN(0, \ I_l \otimes Y_{22})$$

(iii) The matrix $Y_{11.2} = X_1'M_{X_2}X_1$ with $M_{X_2} = I_\lambda - X_2(X_2'X_2)^{-1}X_2'$. M_{X_2} is an idempotent matrix having only eigenvalues equal to 1 or 0. A Cholesky decomposition of this matrix gives

$$M_{X_2} = P \Lambda P', \quad P'P = I_\lambda, \quad \Lambda = \begin{bmatrix} I_{\lambda-l} & 0 \\ 0 & 0 \end{bmatrix}$$

The orthogonal eigenvectors P are also a base of the random variable space. If Z_1 denotes the representation of X_1 using this base, the pdf of Z_1 in the original base can be constructed using the linear combination property of the matric-variate normal, see Press (1972).

$$Z_1 = P'X_1 \Rightarrow p(Z_1) = f_{Mn}(Z_1| \ 0, \ I_n \otimes P'P) = f_{Mn}(Z_1| \ 0, \ I_n \otimes I_\lambda) \Rightarrow$$

$$Z_1 \sim MN(0, I_n \otimes I_\lambda)$$

$$Y_{11.2} = X_1'M_{X_2}X_1 = Z_1'P'P\Lambda P'PZ_1 = Z_1'\begin{bmatrix} I_{\lambda-l} & 0 \\ 0 & 0 \end{bmatrix}Z_1 = Z_{11}'Z_{11}$$

The pdf of Z_{11} is easily constructed using the decomposition of the matric-variate normal with identity covariance matrix.

$$p(Z_{11}) = f_{Mn}(Z_{11}| \ 0, \ I_n \otimes I_{\lambda-l}) \Rightarrow Z_{11} \sim MN(0, \ I_n \otimes I_{\lambda-l})$$

Application of the definition of the Wishart distribution enables one to derive

$$p(Y_{11.2}) = p(Z_{11}'Z_{11}) = f_W(Y_{11.2}| \ I_n, \ \lambda-l) \Rightarrow Y_{11.2} \sim W(I_n, \ \lambda-l)$$

We note that the derivation stated above is the matric–variate analog of the univariate case described by, e.g., Judge et. al. (1985, ch.2) (the pdf of the estimator of the sum of squared residuals in the standard linear model).

Consider the case where $n=1$, $l=m-1$ and $\lambda>m-1$, this decomposition then results in

$$p(Y_{22}) \;=\; f_W(Y_{22}|\ I_{m-1},\lambda)$$

$$p(Y_{21}|Y_{22}) \;=\; f_{Mn}(Y_{21}|\ 0,\ 1{\otimes}Y_{22}) \;\Rightarrow\; y_{21}|Y_{22} \sim n(0,\ Y_{22}) \;\Rightarrow\;$$

$$z \;=\; Y_{22}^{-\frac{1}{2}}y_{21} \;\Rightarrow\; z \sim n(0,I_{m-1})$$

$$p(y_{11.2}) \;=\; f_W(y_{11.2}|1,\lambda-m+1) \;=\; f_{\chi^2}(y_{11.2}|\lambda-m+1) \;\Rightarrow\; y_{11.2} \sim \chi^2(\lambda-m+1)$$

The pdf of Y is characterised by the marginal and conditional pdf's stated above.

By generating Y_{22}, z and $y_{11.2}$ all elements of Y can be calculated recursively. The calculation of the elements of Y can be done efficiently using the following decomposition

$$Y=\begin{bmatrix} y_{11} & Y_{12} \\ Y_{21} & Y_{22} \end{bmatrix} = \begin{bmatrix} y_{11.2}+Y_{12}Y_{22}^{-1}Y_{21} & (Y_{22}^{\frac{1}{2}}z)' \\ Y_{2}^{\frac{1}{2}}\,_2 z & Y_{22} \end{bmatrix} = \begin{bmatrix} y_{11.2}+z'Y_{22}^{\frac{1}{2}\prime}Y_{22}^{-1}Y_{22}^{\frac{1}{2}}z & (Y_{22}^{\frac{1}{2}}z)' \\ Y_{2}^{\frac{1}{2}}\,_2 z & Y_{22}^{\frac{1}{2}\prime}Y_{22}^{\frac{1}{2}} \end{bmatrix}$$

$$=\begin{bmatrix} y_{11.2}+z'z & (Y_{22}^{\frac{1}{2}}z)' \\ Y_{22}^{\frac{1}{2}}z & Y_{22}^{\frac{1}{2}\prime}Y_{22}^{\frac{1}{2}} \end{bmatrix} = \begin{bmatrix} y_{11.2}^{\frac{1}{2}} & 0 \\ z & Y_{22}^{\frac{1}{2}} \end{bmatrix}'\begin{bmatrix} y_{11.2}^{\frac{1}{2}} & 0 \\ z & Y_{22}^{\frac{1}{2}} \end{bmatrix} = U_m'U_m$$

$$U_m=\begin{bmatrix} y_{11.2}^{\frac{1}{2}} & 0 \\ z & Y_{22}^{\frac{1}{2}} \end{bmatrix} = \begin{bmatrix} y_{11.2}^{\frac{1}{2}} & 0 \\ z & U_{m-1} \end{bmatrix}$$

The construction of $Y_{22}^{\frac{1}{2}}$ as a lower triangular matrix in U_{m-1} implies that the decomposition can be applied recursively.

For generating matric–variate t drawings, figure 1 indicates that the square root of an inverted–Wishart random matrix is needed. The lower triangular matrix U_m is equal to the inverse of such a matrix. In figure 2 a flow diagram is shown for the computation of square roots of Wishart distributed matrices. We note that Wishart distributed drawings can be obtained as inner–products of the U_m drawings. A non–standard Wishart can be constructed through a scale transformation, see Anderson (1984). This algorithm for generating Wishart matrices has been used by Geweke (1988).

We end the discussion of generating Wishart drawings with a remark on the relative efficiency of the different procedures. The number of random variab-

Figure 2. Algorithm for generating a square root of a Wishart matrix.

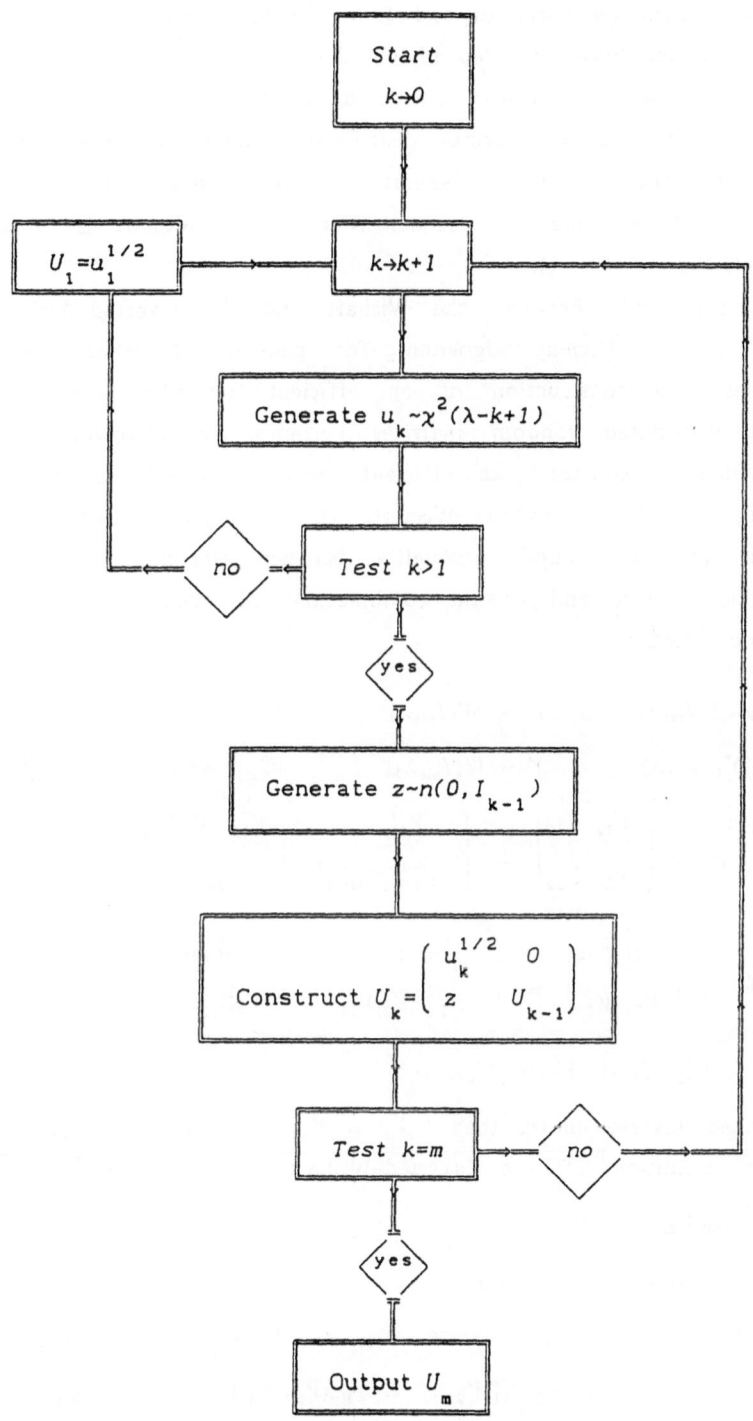

les to be generated for one Wishart distributed random variable is equal to $\frac{1}{2}m(m+1)$ (m χ^2 distributed variables and $\frac{1}{2}m(m-1)$ univariate normal distributed variables). This number does not depend on the degrees of freedom parameter λ. The method of drawing Wishart distributed random matrices by taking the innerproducts of matric-variate normal distributed random matrices needs λm univariate random variables to be generated ($\lambda \times m$ univariate normal random drawings). The algorithm using the decomposition of the Wishart is as a consequence more efficient ($\lambda > m - 1$).

Given the relationship between the Wishart and the inverted-Wishart and the existence of an efficient algorithm for generating Wishart distributed random variables, the construction of an efficient algorithm for generating inverted-Wishart distributed random matrices becomes straightforward. However it is also possible to construct an efficient algorithm based on the decomposition of the pdf of the inverted-Wishart. The decomposition of the inverted-Wishart pdf uses the implicit equality between elements of a Wishart distributed random matrix and certain combinations of elements of an inverted-Wishart random matrix

$$p(V) = f_{iW}(V|\ I_m, \lambda) \qquad V \sim iW(I_m, \lambda) \qquad Y = V^{-1} \Rightarrow$$

$$p(Y) = f_W(Y|\ I_m, \lambda) \qquad Y \sim W(I_m, \lambda) \qquad V_{22.1} = V_{22} - V_{21}V_{11}^{-1}V_{12}$$

$$V = Y^{-1} \Rightarrow Y = \begin{bmatrix} Y_{11} & Y_{12} \\ Y_{21} & Y_{22} \end{bmatrix} = \begin{bmatrix} V_{11.2}^{-1} & -V_{11.2}^{-1}V_{12}V_{22}^{-1} \\ -V_{22.1}^{-1}V_{21}V_{11}^{-1} & V_{22.1}^{-1} \end{bmatrix},$$

$$V = \begin{bmatrix} V_{11} & V_{12} \\ V_{21} & V_{22} \end{bmatrix} = \begin{bmatrix} Y_{11.2}^{-1} & -Y_{11.2}^{-1}Y_{12}Y_{22}^{-1} \\ -Y_{22.1}^{-1}Y_{21}Y_{11}^{-1} & Y_{22.1}^{-1} \end{bmatrix}$$

$-p(Y_{22}) = f_W(Y_{22}|\ I_l, \lambda), \ Y_{22} \sim W(I_l, \lambda)$

The partitioned inverse shows that $V_{22.1}^{-1} = Y_{22}$, so $p(V_{22.1}^{-1}) = f_W(V_{22.1}^{-1}|\ I_l, \lambda)$. The definition of the inverted-Wishart gives

$p(V_{22.1}) = f_{iW}(V_{22.1}|I_l, \lambda).$

$-p(Y_{21}|Y_{22}) = f_{Mn}(Y_{21}|\ 0,\ I_l \otimes Y_{22})$

According to the partitioned inverse, $Y_{21} = -V_{22.1}^{-1}V_{21}V_{11}^{-1}$ and $Y_{22} = V_{22.1}^{-1}$ resulting in $p(V_{22.1}^{-1}V_{21}V_{11}^{-1}|V_{22.1}) = f_{Mn}(V_{22.1}^{-1}V_{21}V_{11}^{-1}|\ 0,\ I_l \otimes V_{22.1}^{-1})$. One conditionalises on $V_{22.1}$ but $V_{22.1}$ also appears in the "random expression". $V_{22.1}$ can be removed from the random expression leading to a change in the variance (mean $=0$) resulting in

$$p(V_{21}V_{11}^{-1}|V_{22.1}) = f_{Mn}(V_{21}V_{11}^{-1}|\ 0,\ I_l \otimes V_{22.1})$$

$$-p(Y_{11.2}) = f_W(Y_{11.2}|\ I_n, \lambda - l)$$

$Y_{11.2} = V_{11}^{-1}$ as is shown by the partioned inverse. So $p(V_{11}^{-1}) = f_W(V_{11}^{-1}|\ I_n, \lambda - l)$. Using the definition of inverted-Wishart random matrices gives:

$$p(V_{11}) = f_{iW}(V_{11}|\ I_n, \lambda - l)$$

For $n=1$, $l=m-1$ and $\lambda > m-1$, the decomposition amounts to

$$p(V_{22.1}) = f_{iW}(V_{22.1}|I_{m-1}, \lambda)$$

$$p(V_{21}v_{11}^{-1}|V_{22.1}) = f_{Mn}(V_{21}v_{11}^{-1}|\ 0,\ I_{m-1} \otimes V_{22.1}) \Rightarrow v_{21}v_{11}^{-1}|V_{22.1} \sim n(0, V_{22.1})$$

$$\Rightarrow z = V_{22.1}^{-\frac{1}{2}}v_{21}v_{11}^{-1} \Rightarrow z \sim n(0, I_{m-1})$$

$$p(v_{11}) = f_{iW}(v_{11}|\ 1, \lambda - m + 1)$$

$$= f_{inv-gam-2}(v_{11}|\ 1, \lambda - m + 1) \Rightarrow v_{11}^{-1} \sim \chi^2(\lambda - m + 1)$$

After generating the random variables z, v_{11} and $V_{22.1}$, the matrix V can be constructed. An elegant and efficient way of construction of V is as follows

$$V = \begin{bmatrix} v_{11} & V_{12} \\ V_{21} & V_{22} \end{bmatrix} = \begin{bmatrix} v_{11}^{\frac{1}{2}}v_{11}^{\frac{1}{2}} & (v_{11}^{\frac{1}{2}}v_{11}^{\frac{1}{2}}V_{22.1}^{\frac{1}{2}}z)' \\ v_{11}^{\frac{1}{2}}v_{11}^{\frac{1}{2}}V_{22.1}^{\frac{1}{2}}z & V_{22.1}+V_{21}v_{11}^{-1}V_{12} \end{bmatrix}$$

$$= \begin{bmatrix} v_{11}^{\frac{1}{2}}v_{11}^{\frac{1}{2}} & (v_{11}^{\frac{1}{2}}v_{11}^{\frac{1}{2}}V_{22.1}^{\frac{1}{2}}z)' \\ v_{11}^{\frac{1}{2}}v_{11}^{\frac{1}{2}}V_{22.1}^{\frac{1}{2}}z & V_{22.1}+(V_{22.1}^{\frac{1}{2}}v_{11}^{\frac{1}{2}}z)(V_{22.1}^{\frac{1}{2}}v_{11}^{\frac{1}{2}}z)' \end{bmatrix}$$

$$= \begin{bmatrix} v_{11}^{\frac{1}{2}} & v_{11}^{\frac{1}{2}}V_{22.1}^{\frac{1}{2}}z \\ 0 & V_{22.1}^{\frac{1}{2}} \end{bmatrix}' \begin{bmatrix} v_{11}^{\frac{1}{2}} & v_{11}^{\frac{1}{2}}V_{22.1}^{\frac{1}{2}}z \\ 0 & V_{22.1}^{\frac{1}{2}} \end{bmatrix} = U'_m U_m$$

$$U_m = \begin{bmatrix} v_{11}^{\frac{1}{2}} & v_{11}^{\frac{1}{2}}V_{22.1}^{\frac{1}{2}}z \\ 0 & V_{22.1}^{\frac{1}{2}} \end{bmatrix} = \begin{bmatrix} v_{11}^{\frac{1}{2}} & v_{11}^{\frac{1}{2}}U_{m-1}z \\ 0 & U_{m-1} \end{bmatrix}$$

When $V_{22.1}^{\frac{1}{2}}$ is constructed as an upper triangular matrix, the decomposition can be applied recursively. An algorithm for generating square roots of inverted-Wishart matrices can be constructed using this decomposition, see Zellner, Bauwens and Van Dijk (1988).

It is of course also possible as suggested earlier, to generate the square roots of the inverted-Wishart random matrices as inverses of square roots of Wishart distributed random matrices. Inverses of lower triangular matrices can be calculated efficiently. Inversion of the lower triangular matrices gives an algorithm for generating the square roots of inverted-Wishart random matrices that is equivalent in computational efficiency compared with the algorithm for generating square roots of inverted-Wishart matrices which we described above.

Figure 3 indicates the changes in the flow diagram of figure 2 that are needed to obtain an algorithm for generating square roots of inverted-Wishart distributed random matrices, which are sufficient to construct matric-variate t drawings. Inverted-Wishart drawings can be constructed by taking the inner products of the square roots. A scale transformation gives non-standard inverted-Wishart distributed random matrices.

Figure 3. Changed blocks in the flow diagram in figure 2 for generating a square root of an inverted-Wishart matrix

$$\boxed{\text{Generate } u_k \sim inv\text{-}\chi^2(\lambda-k+1)}$$

$$\boxed{\text{Construct } U_k = \begin{pmatrix} u_k^{1/2} & u_k^{1/2} z' U_{k-1} \\ 0 & U_{k-1} \end{pmatrix}}$$

4. Structural form analysis using reduced form drawings from a matric-variate t

The marginal posterior density of the parameters of a multivariate regression model is matric-variate t using a diffuse prior, see Zellner (1971). The reduced form of exactly identified Simultaneous Equation Models (SEM) belongs to this class of multivariate regression models. The reduced form of a SEM can be derived as follows. The standard form of the SEM is

$$YB + Z\Gamma = U \Leftrightarrow (Y\ Z) \begin{bmatrix} B \\ \Gamma \end{bmatrix} = U$$

The following assumptions are usually made concerning the SEM model (Judge et.al. (1985)):

$-Y$: $T \times m$ matrix of endogenous variables with rank m

-Z : $T \times k$ matrix of predetermined variables with rank k, $(Y\ Z)$ rank $k+m$

-U : $T \times m$ matrix of unobserved disturbances

-"orthogonality" of Z with respect to U : $\forall t \leq T$, $\forall i \geq 0$ such that $t+i \leq T$, the t-th row of Z is assumed to be independent of the $(t+i)$-th row of U

-$\begin{bmatrix} B \\ \Gamma \end{bmatrix}$ has full rank m

-$U \sim MN(0\ , \Sigma \otimes I_T)$

The implied reduced form of the SEM can be constructed by multiplying the SEM by the inverse of B

$$Y = -Z\Gamma B^{-1} + UB^{-1} \Leftrightarrow Y = Z\Pi(\Gamma, B) + V(B)$$

If the SEM is exactly identified, the relationship between the matrices Π and B, Γ is one-to-one. This implies that there are no restrictions placed on the reduced form parameters Π. The reduced form belongs to the class of multivariate regression models when the model is exactly identified. Given a diffuse prior, the parameters of the reduced form have a matric-variate t distribution. This result and the algorithms of section 3 enable us to generate reduced form parameters. Through the one-to-one correspondence between the parameters of the reduced form, Π, and the structural form, B and Γ, it is also possible to generate structural form parameters. For expository purposes use is made of two different supply and demand models describing the behaviour of prices and quantities of an agricultural product (meat). First the so called "Tintner model" describing the US meat market, Tintner (1952), and second the "Belgian Beaf Market model" (BBM) from Morales (1971). These two meat market models have the following specification

$$q = \beta_1 p + \gamma_1 y + u_1 \qquad \text{Supply equation}$$

$$q = \beta_2 p + \gamma_2 c + u_2 \qquad \text{Demand equation}$$

with :- q - quantity of meat consumed per capita
- p - price of meat
- c - cost of processing meat (Tintner), cattle stock per capita (Morales)
- y - national income per capita

The models attempt to explain the price and quantity behaviour in the meat market given national income per capita and a certain cost factor for processing meat. The series p and q are therefore assumed to be endogenous and the series y and c exogenous. (Note that in the original model also constant terms are incorporated. By treating the series in deviations from their means, these constants are deleted from the model.) The meat market model has the following standard specification

$$Y = (q\ p) \qquad\qquad Z = (c\ y) \qquad\qquad U = (u_1\ u_2)$$

$$B = \begin{bmatrix} 1 & 1 \\ -\beta_1 & -\beta_2 \end{bmatrix} \qquad\qquad \Gamma = \begin{bmatrix} -\gamma_1 & 0 \\ 0 & -\gamma_2 \end{bmatrix}$$

From the definition of B and Γ, it is possible to derive the implied reduced form matrix Π

$$\Pi = -\Gamma B^{-1} = (\beta_1 - \beta_2)^{-1} \begin{bmatrix} -\beta_2\gamma_1 & -\gamma_1 \\ \beta_1\gamma_2 & \gamma_2 \end{bmatrix} = \begin{bmatrix} \pi_{11} & \pi_{12} \\ \pi_{21} & \pi_{22} \end{bmatrix}$$

The structural form parameters can be obtained from the reduced form parameters

$$\beta_1 = \pi_{21}/\pi_{22} \qquad\qquad\qquad \beta_2 = \pi_{11}/\pi_{12}$$

$$\gamma_1 = \pi_{11} - \pi_{12}\pi_{21}/\pi_{22} \qquad\qquad \gamma_2 = \pi_{21} - \pi_{11}\pi_{22}/\pi_{12}$$

These structural form parameters are ratios of reduced form parameters. When the denominators of these ratios become 0, the structural form parameters become in general infinite. The reduced form parameters have a matric-variate t distribution in the case of the meat market models. These matric-variate t densities have a non-zero value at 0. However being only a point in the parameterspace, the probability that the denominators are equal to 0 is 0. The resulting joint density of the structural form parameters will converge to 0 rather slow though resulting in a finite distribution but infinite mean and variances, see Kleibergen and Van Dijk (1992). Note that this density is still only an approximation of the posterior density of the structural parameters : weights have to assigned to each parameter drawing to make the posterior of the structural parameters proportional to the likelihood. These weights are equal to the jacobian of the transformation from B, Γ to Π and for these models are equal to

$$\left| \frac{\partial(B\ \Gamma)}{\partial\Pi} \right| = \left| \begin{bmatrix} 0 & \pi_{12}^{-1} & 1 & -\pi_{22}\pi_{12}^{-1} \\ \pi_{22}^{-1} & 0 & -\pi_{12}\pi_{22}^{-1} & 1 \\ 0 & -\pi_{11}\pi_{12}^{-2} & -\pi_{21}\pi_{22}^{-1} & \pi_{11}\pi_{22}\pi_{12}^{-2} \\ -\pi_{21}\pi_{22}^{-2} & 0 & \pi_{12}\pi_{21}\pi_{22}^{-2} & -\pi_{11}\pi_{12}^{-1} \end{bmatrix} \right|$$

$$= \left| \pi_{12}^{-1} \right| \left| \pi_{22}^{-1} \right| \left| (\pi_{11}^{-1}\pi_{12}^{-1} - \pi_{21}^{-1}\pi_{22}^{-1})^2 \right|$$

$$= |\beta_1 - \beta_2|^4 |\gamma_1\gamma_2|^{-1}$$

In general the jacobian is an increasing function in the structural form parameters which correspond to the endogenous variables. This results in even thicker tails which results in infinite posterior moments, see Kleibergen and Van Dijk (1992). So we truncated the parameter region in our applications.

The method of generating reduced form parameters is applied to the Tintner and BBM meat market models using the SISAM program, see Hop and Van Dijk (1992). The bivariate posteriors of (β_1,β_2), (β_1,γ_1) and (β_2,γ_2) for the Tintner model are shown in the figures 1-3. The posteriors in these figures are not completely smooth because of the not complete convergence towards the limiting functions but show the specific features of the posteriors quite well. The bivariate posteriors are very skewed because the behaviour of the posterior in different directions differs substantially. This confirms the earlier statement about the probability of high values of the structural form parameters. For a theoretical analysis of this problem reference is made to Kleibergen and Van Dijk (1992).

Figure 4. bivariate posterior of (β_1,β_2) in the Tintner meat market model.

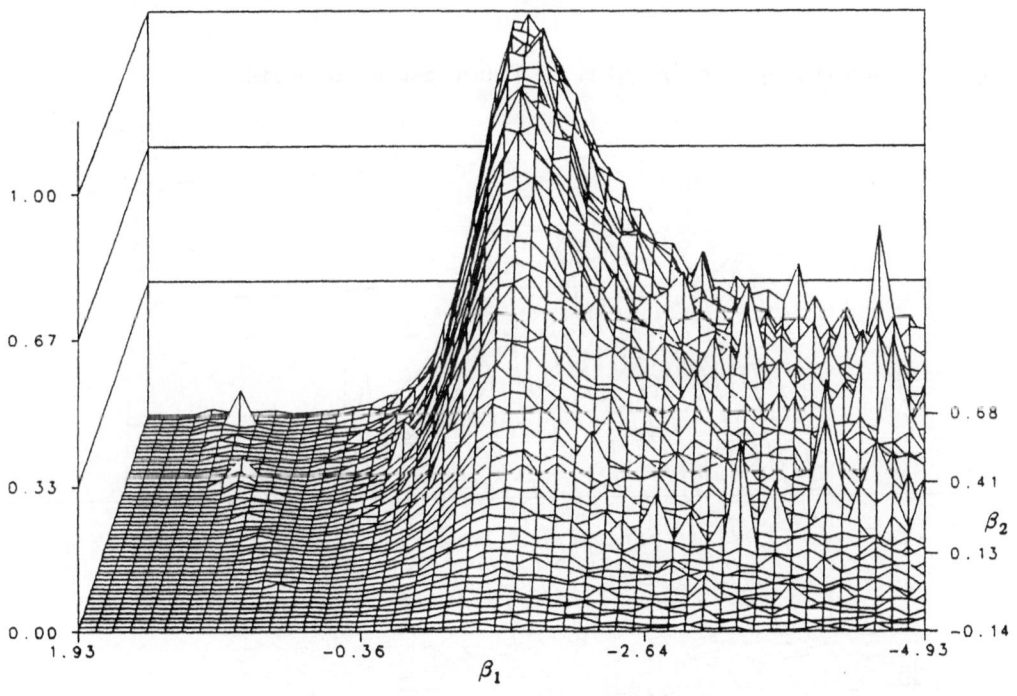

44

Figure 5. bivariate posterior of (β_1, γ_1) in the Tintner meat market model.

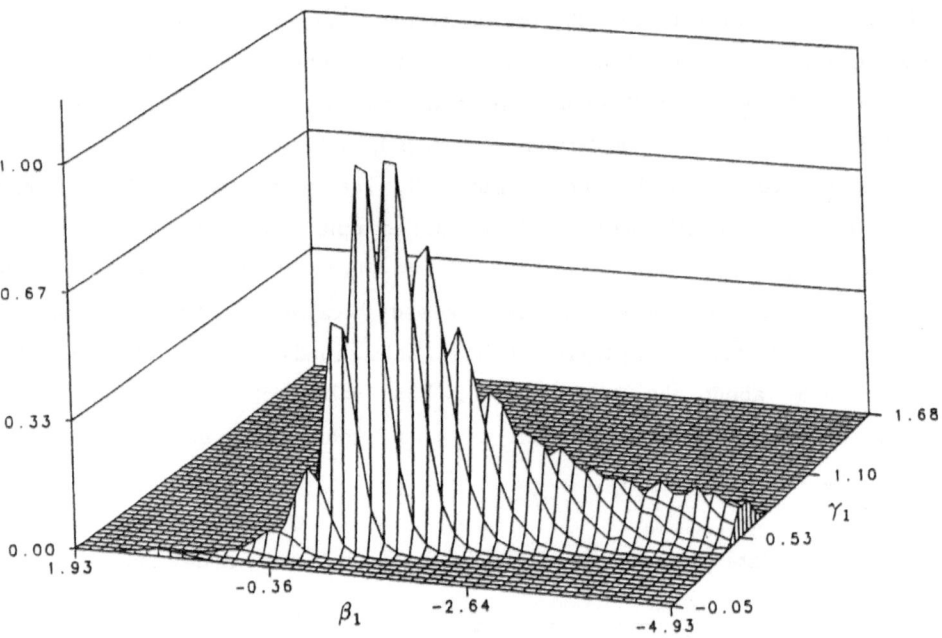

Figure 6. bivariate posterior of (β_2, γ_2) in the Tintner meat market model.

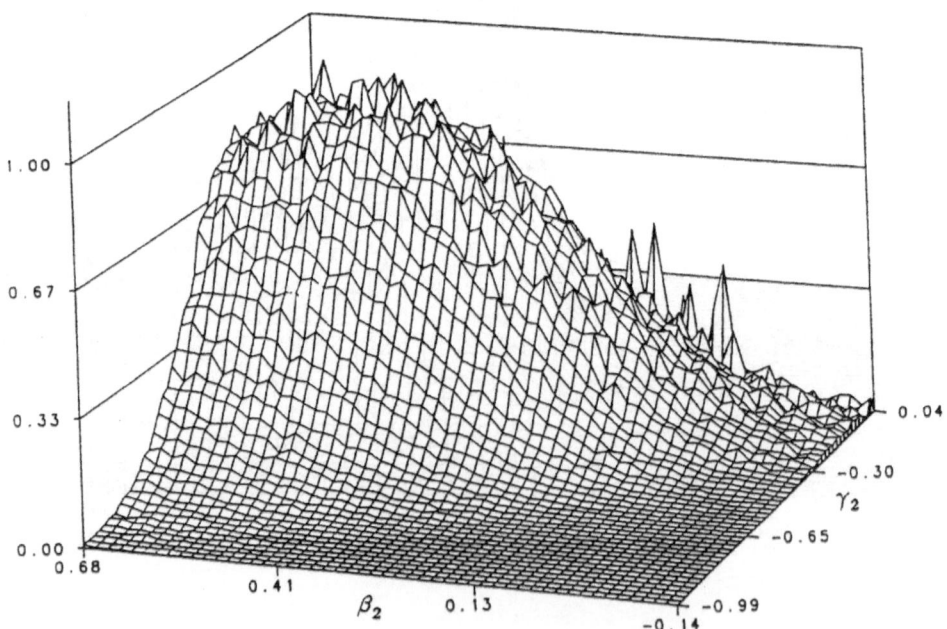

The posterior means and standard deviations of the structural form parameters (defined on a truncated parameter region) and the classical estimates of these parameters of both meat market models are shown in table 1.

Table 1. Posterior means and classical estimates (ILS) of the parameters of the Tintner and BBM models (stand. dev. below)

	β_1	β_2	γ_1	γ_2
Tintner :				
mean	0.42	−2.30	0.28	−0.28
	0.18	1.18	0.13	0.17
ILS	0.62	-1.42	0.19	-0.42
	0.36	0.54	0.06	0.22
BBM :				
mean	0.18	−0.84	0.81	1.06
	0.07	0.27	0.15	0.15
ILS	0.14	−0.69	0.73	1.11
	0.05	0.12	0.07	0.11

The differences between the posterior means of the structural form parameters and the classical estimates is quite substantial for the Tintner model. The posterior means and estimates for the BBM model are more or less the same. Classical statistical analysis makes use of asymptotic theory. Given the differences between the means and estimates for the Tintner model, the applicability of asymptotic theory is doubtful. The shape of the posteriors of the structural form parameters indicates that the posteriors are highly non-normal.

5. Conclusions

Two algorithms for generating random matrices with (inverted-) Wishart and matric-variate t distributions are presented. These algorithms are derived using formal distribution theory and the derivation turns out to be quite simple. The algorithms are used to generate matric-variate t distributed reduced form parameters of SEM's. Two exactly identified meat market models are used in this example because the distribution of the reduced form parameters is matric-variate t for exactly identified models. Special weights are attached to each random drawing of the reduced form parameters to assure that the joint posterior of the structural form parameters is proportional to the SEM likelihood. These weights have a formal theoretic ground and equal the jacobian of the transformation from structural form to reduced form. The resulting posteriors of the structural form do not confirm very well with the asymptotic normality of the pdf's of the classical structural form estimators. As a consequence the application of asymptotic theory seems rather restrictive for several SEM's.

References

- Anderson, T.W., 1984, *An introduction to multivariate statistical analysis*, Wiley, New York
- Bauwens, L., 1984, *Bayesian full information analysis of simultaneous equation models using integration by Monte Carlo*, Berlin, Springer–Verlag
- Box, G.E.P. and G.C. Tiao, 1973, *Bayesian inference in statistical analysis*, Addison–Wesley, Reading, MA
- DeGroot, M., 1987, *Probability and Statistics*, 2nd edn, Addison–Wesley, Reading, MA
- DeJong, D.N. and C.H. Whiteman, 1991, *Trends and Random Walks in Macro–Economic Time Series : a Reconsideration based on the Likelihood Principle*, Journal of Monetary Economics, forthcoming
- Drèze, J.H. and J.F. Richard, 1983, *Bayesian analysis of simultaneous equations systems*, in: Z. Griliches and M.D. Intrilligator, eds., Handbook of Econometrics, Vol. 1., North–Holland Publishing Co., Amsterdam
- Geweke, J., 1986, *Exact inference in the Inequality Constrained Normal Linear Regression Model*, Journal of Applied Econometrics, 1, 127–141
- Geweke, J., 1988, *Antithetic Acceleration of Monte–Carlo Integration in Bayesian Inference*, Journal of Econometrics, 38, 73–90
- Hogg, R.V. and A.T. Craig, 1978, *Introduction to Mathematical Statistics*, 4–th edn, Macmillan, New York
- Hop, J.P. and H.K. van Dijk, 1992, *SISAM and MIXIN : two algorithms for the computation of posterior moments and densities using Monte–Carlo integration*, Computer Science in Economics and Management, forthcoming
- Judge, G.G, W.E. Griffiths, R.C. Hill, H. Lütkepohl and T.C. Lee, 1985, *The Theory and Practice of Econometrics*, 2nd edn, Wiley, New–York
- Kinderman, A.J. and J.F. Monahan, 1980, *New methods for Generating Student t and Gamma Variables*, Computing, 25, 369–377
- Kleibergen, F.R. and H.K van Dijk, 1992, *Bayesian Simulateneous Equations Model analysis : On the existence of structural posterior moments*, Working paper, Econometric Institute, Erasmus University Rotterdam
- Morales, J.A., 1971, *Bayesian Full Information Structural Analysis*, Berlin, Springer–Verlag
- Press, S.J., 1972, *Applied multivariate analysis*, Rinehart and Winston, New York
- Raiffa, H. and R. Schlaiffer, 1961, *Applied statistical decision theory*, Graduate School of Business Administration, Harvard University, Boston
- RATS 3.0, *VAR econometrics*, Evanston, Illinois
- Tintner, G., 1952, *Econometrics*, Wiley, New–York
- Van Dijk, H.K. and T. Kloek, 1980, *Further experience in Bayesian analysis using Monte–Carlo integration*, Journal of Econometrics 14, 307–328
- Zellner, A., 1971, *An Introduction to Bayesian Inference in Econometrics*, Wiley, New–York
- Zellner, A., L. Bauwens and H.K. van Dijk, 1988, *Bayesian specification analysis and estimation of simultaneous equation models using Monte–Carlo integration*, Journal of Econometrics, 38, 39–72

Approximate HPD Regions for Testing Residual Autocorrelation Using Augmented Regressions

L. Bauwens

CORE, Université Catholique de Louvain,

34 Voie du Roman Pays, 1348 Louvain-La-Neuve, BELGIUM .

A. Rasquero

GREQE, Ecole des Hautes Etudes en Sciences Sociales,

2 Rue de la Charité, 13002 Marseille, FRANCE

Key Words: Residual Autocorrelation, HPD Region, Power, Augmented Regressions, Regression Model, Bayesian Inference.

Abstract

We evaluate two tests of residual autocorrelation in the linear regression model in a Bayesian framework. Each test checks if an approximate highest posterior density region of the parameters of the autoregressive process of the error contains the null hypothesis. The approximation consists in computing the posterior density of the coefficients of the AR process using augmented regressions. The first test uses the initial regression augmented with its lagged Bayesian residuals and can be done with tables of the Fisher distribution. The second test augments the initial regression with lagged dependent and explanatory variables, and requires numerical integration. The tests are evaluated through a small Monte-Carlo experiment, which indicates that the first test (easier to compute) is more powerful than the second one.

1. Introduction

We propose and evaluate two tests of residual autocorrelation in linear regression models. The tests are based on approximate highest posterior density (HPD) regions for the parameters of an autoregressive (AR) process on the error term of the regression. The evaluation is done using data generated with a simple known model; in particular we compute the power of the tests although they have a Bayesian interpretation.

Residual autocorrelation means that an important aspect of the model has been misspecified, as this deficiency may be caused by omitted variables, the inadequacy of the functional form, or unsufficient dynamics. It may have severe effects. From the economic viewpoint, it leads to a wrong interpretation of the phenomenon under study. From the statistical standpoint, it is in conflict with the generally done hypothesis of independently and identically distributed error terms; hence the posterior and predictive densities may be misleading, or from a classical perspective, the estimation technique used may be inefficient. For those reasons, a test of autocorrelation must be a stage of the econometric modelling of time series.

Several frequentist tests have been available for many years, from the traditional Durbin-Watson statistic to the score test which can easily be implemented by augmenting the initial regression with its lagged residuals and testing their joint significance. In Bayesian econometrics, the research on this subject is more recent. Drèze and Mouchart (1990) propose to use the classical Durbin-Watson statistic and to examine the plot of the residuals as a quick test, but they recognize that more formal Bayesian procedures for testing autocorrelation could be useful. The test proposed by Aprahamian, Lubrano, and Marimoutou (1990) is based on a measure of divergence between the posterior densities of the parameters of interest under the null hypothesis (no autocorrelation) and under the alternative one (autocorrelation), the latter being implemented through the same augmented regression as a classical score test. These authors apply a general principle proposed by Hausman (1978) and extended by Florens and Mouchart (1989) to the Bayesian paradigm, but they are interested in the influence of misspecification on the parameters of interest rather than in misspecification itself. A problem with this approach is that the parameters of interest are not necessarily the same under the null hypothesis and under the alternative one. For example, if consumption is regressed on income in a static model, one finds usually *residual* autocorrelation. Extending the same static model with autocorrelated *errors* is not very interesting as it implies that the short run and the long run propensities to consume are equal. An often better solution extends the static model to a dynamic one by adding lagged consumption and income as regressors. Then one can have different short run and long run propensities and even test their equality; see the paper of Hendry and Mizon (1978).

An ideal way to test autocorrelation is to compute the *exact* posterior density of α, the vector of coefficients of the AR(p) process of the errors. The decision is to reject the null hypothesis ($\alpha = 0$) if an HPD region of preselected probability does not contain the point $\alpha = 0$. We do not consider using posterior odds because the null hypothesis is a single point in the parameter space and we consider it is arbitrary to assign a non-zero prior probability to a point; see Poirier (1987) for a posterior odds approach. The posterior density of α has been derived by Zellner (1971) for the case p = 1, and by Richard (1977) for the general case, but it does not belong to a known family (such as normal, Student-t, poly-t...). The computation of an exact HPD region is then a very difficult problem; even an approximate region based on the ellipsoid of concentration is not easy to obtain since the posterior moments of α must be computed by Monte-Carlo integration (using either importance sampling or Gibbs sampling). Hence the ideal test would be difficult to implement in practice: it would require very heavy computations; even if these were programmed once for all, they would not be useful in an applied study where one wants a quick diagnostic, often several times.

To achieve this goal, we employ two kinds of augmented regressions that provide an approximate posterior density of α which is used to build an approximate HPD region for testing the nullity of α. The approximate density could be used also as an importance function for the computation of the exact density of α and of its moments if an ideal test is desired, but this issue is not pursued in this paper. We investigate instead if the approximate posterior density gives a good idea of the value of α and provides a safe diagnostic on autocorrelation.

Section 2 reviews the linear regression model with autocorrelated errors, partly to set the notations. Section 3 covers the first approximate test, which is based on the initial regression augmented with its Bayesian residuals. Section 4 explains the two versions of the second test, which comes from augmenting the initial regression with lags of the dependent and explanatory variables. Section 5 provides the Monte-Carlo results and Section 6 concludes.

2. The normal linear regression model with autocorrelated errors

Consider the linear regression model

(2.1) $$y_t = x_t'\beta + u_t \qquad (t = 1...T),$$

or in matrix format

(2.2) $$y = X\beta + u$$

where y is the $T \times 1$ endogenous variable, $y = (y_1,...,y_T)'$,

$X = (x_1,...,x_T)'$ is the $T \times k$ matrix of weakly exogenous regressors,

β is the $k \times 1$ vector of parameters, $\beta = (\beta_1,....,\beta_k)'$, and

$u = (u_1,...,u_T)'$ is the $T \times 1$ vector of errors.

Under H_0, we assume $u \sim N_T(0,\sigma^2 I_T)$.

Under H_1, the errors are supposed to follow an AR(p) Gaussian process defined by

(2.3) $$A(L)u_t = u_t - \alpha_1 u_{t-1} - ... - \alpha_p u_{t-p} = \varepsilon_t \qquad t = 1...T$$

where L is the lag operator; in matrix format (2.3) becomes

(2.4) $A(L)u = \varepsilon$

where $\varepsilon = (\varepsilon_1, \ldots, \varepsilon_T)' \sim N_T(0, \sigma^2 I_T)$. We define $\alpha = (\alpha_1, \ldots, \alpha_p)'$.

In the model with autocorrelated errors, we assume that p initial conditions (indexed from time 1-p to 0) on y_t and X_t are given, which corresponds to the widespread practice of keeping the first p observations for this purpose. For some statistical procedures (such as the comparison of properties of estimators of β in repeated sampling, and encompassing tests) both models must share the same sample space: it suffices then to assume that the observations from 1-p to 0 are not used under the null hypothesis. Our procedures do not rely on such an assumption and are in the spirit of a Wald approach to testing: estimation is done under the alternative hypothesis and a parameter restriction is tested. Moreover, by conditioning on initial values, no restriction has to be put on the roots of A(L), such as the exclusion of unit roots, since the expression of the likelihood function does not depend on the order of magnitude of the roots of A(L).

Bayesian inference on the normal linear model under H_0 is well-known, see eg Zellner (1971). Under H_1, the inference is not so easy since the substitution of (2.1) in (2.3) shows clearly that the model is non-linear in the parameters:

(2.5) $A(L)y_t = A(L)x_t'\beta + \varepsilon_t$

The likelihood function is proportional to the Normal data density (which is implicitly conditional on the initial p observations and on X)

(2.6) $y_\alpha | \beta, \alpha, \sigma^2 \sim N_T(X_\alpha \beta, \sigma^2 I_T)$

where $y_\alpha = A(L)y$ and $X_\alpha = A(L)X$.

We take the prior density of β and σ^2 as Normal-Inverted Gamma (the conjugate prior for the likelihood under H_0), namely

(2.7) $\beta | \sigma^2 \sim N_k(\beta_0, \sigma^2 M_0^{-1})$ and $\sigma^2 \sim IG(s_0^2, \nu_0)$.

The prior $p(\alpha)$ is non-informative as α is in our approach a nuisance parameter. For simplicity, we take

(2.8) $p(\alpha) \propto 1, \ \alpha \in \mathbb{R}^p,$

although it could easily be truncated to a subspace of $\mathbb{R}^p$, eg for imposing that A(L) has all its roots outside the unit circle.

The posterior density of β conditional on α is the Student density

(2.9) $\beta | \alpha, y \sim t_k(\beta_*, M_*, s_*^2, \nu_*)$

and given (2.8), the marginal posterior density of α is

(2.10) $p(\alpha | y) \propto [(s_*^2)^{-\nu_*} |M_*|]^{1/2}$

where $M_* = M_0 + X'_\alpha X_\alpha$, $\beta_* = M_*^{-1}(X'_\alpha y_\alpha + M_0 \beta_0)$, $\nu_* = \nu_0 + T$, and

$$s_*^2 = s_0^2 + (y_\alpha - X_\alpha \beta_0)'(I_T + X'_\alpha M_0^{-1} X_\alpha)^{-1}(y_\alpha - X_\alpha \beta_0).$$

The normalizing constant and the moments of $p(\alpha|y)$ are not known analytically and must be computed by numerical integration. Importance sampling could be used if a good enough approximation of $p(\alpha|y)$ can be built. Richard (1977) proposes to approximate (2.10) by a poly-t density but this suggestion has not been implemented. Other possible importance functions are the approximate posterior densities of α we build in Sections 3 and 4 for testing. An alternative method is to use Gibbs sampling, see Bauwens and Lubrano (1992). The choice of $p(\alpha)$ could be left free in an exact analysis since numerical integration is required.

3. The Bayesian Residuals Test (BRT)

This test was proposed originally by Pagan (1978), for the case of autocorrelation of order one

(3.1) $$u_t = \rho u_{t-1} + \varepsilon_t$$

(using the traditional ρ instead of α_1). The substitution of the right-hand side for u_t in (2.1) gives

(3.2) $$y_t = x'_t \beta + \rho u_{t-1} + \varepsilon_t$$

Since u_{t-1} is not observable, we replace it by the Bayesian residual under H_0,

(3.3) $$\hat{u}_{t-1} = y_{t-1} - x'_{t-1} E(\beta|y),$$

where $E(\beta|y)$ is equal to β_* defined after (2.10) when α is equal to 0. Bauwens and Lubrano (1991) show that if the sample size T tends to infinity, the Bayesian residual $\hat{u}_t$ tends under H_0 to the true (realized) unobservable value of u_t. Relying on this asymptotic argument, we use the posterior density of ρ in

(3.4) $$y_t = x'_t \beta + \rho \hat{u}_{t-1} + \varepsilon_t$$

as an approximation of the posterior density of ρ in (3.1)-(3.2).

In the case of higher order autocorrelation, the generalisation of the previous approach gives in matrix notations

(3.5) $$y = X\beta + Z\alpha + \varepsilon = W\gamma + \varepsilon,$$

where $Z = (\hat{u}_{-1} ... \hat{u}_{-p})$, the $T \times p$ matrix of lags of $\hat{u} = (\hat{u}_1 ... \hat{u}_T)'$.

Setting $M_0 = 0$, $\beta_0 = 0$, $s_0^2 = 0$, $\nu_0 = -k$ in (2.7), one obtains the non-informative prior $p(\beta, \sigma^2, \alpha) \propto \sigma^{-2}$. Applying standard Bayesian formulas, the posterior density of α in (3.5) is given by the Student density

(3.6) $\quad \alpha | y \sim t_p(\hat{\alpha}, Z'M_XZ, y'M_Wy, T-k-p)$

where $\hat{\alpha} = (Z'M_XZ)^{-1}Z'M_Xy$, and $M_Q = I_T - Q(Q'Q)^{-1}Q'$. *We take (3.6) as an approximation of the exact posterior density defined by (2.10)* (specialised to the case where the prior is non-informative). The generalisation of (3.6) to arbitrary values of the prior parameters M_0, β_0, s_0^2 and ν_0 complicates the notations but is standard.

Given (3.6), the HPD region of level π (eg 0.95) for α is the set of points on or in the ellipsoid $\mathcal{E}_\pi$ centered on $\hat{\alpha}$, defined by $p(\alpha|y) = c_\pi$, where c_π is the constant such that $P(\alpha \in \mathcal{E}_\pi|y) = \pi$. If the point 0 corresponding to H_0 is outside $\mathcal{E}_\pi$, H_0 is rejected at the level of confidence π. If $\alpha \sim t_p(a,M,s,\nu)$, then the quadratic form

(3.7) $\quad F(\alpha) = (\alpha-a)'M(\alpha-a).\nu/(ps) \sim F(p,\nu),$

where $F(p,\nu)$ is Fisher's distribution, see Zellner (1971, 383-385). If $F(0)$ is larger than the $(1-\pi)$ quantile of the $F(p,\nu)$ distribution, H_0 is rejected. We call this test **BRT**, the **Bayesian Residuals Test**. When H_0 is rejected, it may be helpful to look at the marginal posterior density of each element of α, in order to detect its elements which are different of 0 (the univariate HPD regions are intervals that can be computed using the quantiles of the univariate t distribution with ν degrees of freedom).

Under the non-informative prior leading to (3.6), BRT is equivalent to a classical score statistic that tests autocorrelation in the model defined by (2.1) and (2.3) through the F-test for $\alpha = 0$ in (3.5); see eg Harvey (1990, 277-278).

The BR test is available in the Bayesian Interactive Program (BIP) described by Aprahamian, Bauwens and Lubrano (1990).

4. The Common Factor Tests (CFST and CFJT)

Under H_1, the linear model (2.1) has a common factor restriction which is obvious in (2.5). If we do not impose the restriction, we get an autoregressive distributed lags (ADL) model wherein we can test the common factor restriction, and conditionally upon acceptation, we can test the hypothesis $\alpha = 0$; this test is called the **Common Factor Sequential Test (CFST)**. We can also test the two restrictions (common factor and $\alpha = 0$) jointly rather than sequentially, hence the **Common Factor Joint Test (CFJT)**.

The ADL model can be written as

(4.1) $\quad A(L)y_t = x_t'B(L) + \varepsilon_t$

where $B(L) = [b_1\ B_2(L)\ ...\ B_k(L)]'$ and $B_j(L) = \sum_0^p b_{jl}L^l$, assuming that the first element of x_t is constant. The model (4.1) has $(k-1)(p+1)+p+1$ coefficients, whereas the model (2.5) has $p+k$ coefficients. There are $(k-1)p$ non-linear restrictions to impose on (4.1) to reduce it to (2.5), namely

(4.2) $b_{j0}\alpha_i + b_{ji} = 0$ (i = 1 to p, j = 2 to k).

We denote by δ the vector of all the non-linear functions defined in (4.2). The common factor hypothesis in (4.1) is $\delta = 0$ (in Section 5, we consider the case where p = 1 and k = 2, so that δ is a scalar). We write (4.1) in matrix format as

(4.3) $y = Y_p\alpha + X\beta + X_p\gamma + \varepsilon = Z\theta + \varepsilon$

where $Z = (Y_p \quad X \quad X_p)$, a T×m matrix, m = p+k+p(k-1),

   $Y_p = (y_{-1} \dots y_{-p})$, the T×p matrix of lags of y,

   $X_p = (X_{-1} \dots X_{-p})$, the T×p(k-1) matrix of lags of the k-1 last columns of X,

   $\theta = (\alpha' \; \beta' \; \gamma')'$, $\beta' = (b_1 \; b_{20} \dots b_{k0})$, $\gamma' = (b_{21} \; b_{31} \dots b_{k1} \; b_{22} \dots b_{kp})$.

For the prior $p(\theta,\sigma^2) \propto \sigma^{-2}$, the posterior density of θ is the Student density

(4.4) $\theta|y \sim t_m(\hat{\theta}, Z'Z, y'M_z y, T-m)$,

where $\hat{\theta} = (Z'Z)^{-1}Z'y$. We assume that T is larger than m+2, for the existence of the posterior variance-covariance matrix of θ. Again, (4.4) can easily be extended to the case where the prior is informative in the natural-conjugate class. We explain how we implement our tests starting from (4.4).

CFJT tests H_0: $\phi' = (\delta' \; \alpha') = 0'$ where ϕ has pk elements. To implement the test, we define a procedure in two steps to check if the point 0 lies outside the HPD region of level π of the parameter space.

i) Since δ is a non-linear function of θ, the posterior density of δ is not known analytically. To compute it, we generate N realizations of θ according to its Student density (4.4) and calculate for each drawing the corresponding value of δ. By simple averaging of these N values and of functions of them we can obtain (if N is large enough) a very good approximation of the posterior expectation $E(\delta|y)$ and of the variance-covariance matrix $V(\delta|y)$. The covariance matrix $Cov(\delta,\alpha|y)$ can be computed likewise. $E(\alpha|y)$ and $V(\alpha|y)$ being known analytically, this step provides us with $E(\phi|y)$ and $V(\phi|y)$ that are used in the next step.

ii) Since ϕ is multivariate, we define the *Pseudo-F* quadratic form

(4.5) $PF(\phi) = K.[\phi-E(\phi|y)]'[V(\phi|y)]^{-1}[\phi-E(\phi|y)]$

where K = (T-m)/[(T-m-2)pk]. If ϕ were distributed as Student with expectation $E(\phi|y)$, variance-covariance matrix $V(\phi|y)$, and T-m degrees of freedom, $PF(\phi)$ would be distributed as F(pk,T-m). Since this is not true, we tabulate the distribution of $PF(\phi)$ by another simulation of (4.4): given M drawings of θ, we can compute M values of ϕ and of $PF(\phi)$, ie for j = 1 to M,

(4.6) $PF(\phi_j) = K.[\phi_j-E(\phi|y)]'[V(\phi|y)]^{-1}[\phi_j-E(\phi|y)]$.

With these M values, we build a histogram which can be smoothed to approximate the posterior density of $PF(\phi)$, and its quantiles, in particular the $(1-\pi)$ quantile. We calculate the value of $PF(\phi)$ under H_0, ie $PF(0)$, and we reject H_0 if $PF(0)$ is larger than the $(1-\pi)$ quantile of the distribution of $PF(\phi)$.

A rejection of H_0 can be interpreted as evidence that there is no common factor ($\delta \neq 0$, with $\alpha = 0$ or $\alpha \neq 0$), or if there is one ($\delta = 0$), that α is different from 0. The alternative hypothesis of this test is therefore more general than the hypothesis of absence of autocorrelation of u_t in (2.1), because of the combination ($\delta \neq 0$, $\alpha = 0$). To avoid this problem which could result in a loss of power of the test, one could perform a sequential test: test $\delta = 0$, and if this is not rejected, test $\alpha = 0$. The first test can be done by using a Pseudo-F for δ, and is an almost free by-product of the procedure of the CFJ test. The second test should be done using the posterior distribution of α *conditioned on* $\delta = 0$, which is nothing else but (2.10), and cannot be computed by simulation from (4.4), because the event $\delta = 0$ is of zero probability; as a (surely imperfect) subsitute we use the marginal Student density of α obtained from (4.4) to test $\alpha = 0$.

The HPD region for ϕ computed through the Pseudo-F test is approximate unless the posterior distribution of ϕ is symmetrical. Some asymmetry in the posterior density of ϕ is caused by the non-linearity of δ with repect to θ. Some limited experiments we have done have revealed a negligible skewness in the density of δ.

5. Monte-Carlo results

We have generated data with a simplified version of (2.1)-(2.3),

(5.1) $$y_t = \beta_1 + \beta_2 x_t + u_t \qquad t=1,\ldots,T$$
(5.2) $$u_t = \rho u_{t-1} + \varepsilon_t \qquad \varepsilon_t \sim N(0,\sigma^2) \, , \ cov(\varepsilon_t,\varepsilon_s)=0 \ \forall t \neq s$$

having the following characteristics:

-the sample size T takes the values 15, 30, 60, and 100;

-x_t is the first difference of the logarithm of an interest rate series in Great Britain and looks like a stationary series;

-ρ takes the values ±0.99, ±0.9, ±0.5, ±0.2, 0;

-$\beta_1 = 0.05$, $\beta_2 = 20$, $\sigma^2 = 0.05$; with these values, the R^2 of equation (5.1) is equal to 0.8 when ρ is equal to 0;

-the prior density is non-informative, being proportional to σ^{-2}.

-ε_t (t = 0 to T) is drawn as a sequence of independent $N(0,\sigma^2)$, u_t is computed by (5.2) starting with $u_0 = (1-\rho^2)^{-1/2}\varepsilon_0$, and then y_t is computed by (5.1). We use the same generated sequence ε_t for each sample size so as to have exactly the same data for a given value of ρ.

With this setup, the augmented regression for the Bayesian residual test is

$$(5.3) \qquad y_t = \beta_1 + \beta_2 x_t + \rho \hat{u}_{t-1} + \varepsilon_t, \qquad t=1,\dots,T.$$

The augmented regression for the common factor tests is the ADL model

$$(5.4) \qquad y_t = \rho y_{t-1} + b_1 + b_2 x_t + \gamma x_{t-1} + \varepsilon_t, \qquad t=1,\dots,T.$$

The common factor restriction is $\delta \equiv \rho b_2 + \gamma = 0$. Substituting $-\rho b_2$ for γ in (5.4), one finds (5.1) and (5.2) by setting $b_2 = \beta_2$ and $b_1 = \beta_1(1-\rho)$

In Table 1, we summarise the results of applying the tests BRT, CFST, and CFJT to each single sample generated for the different pairs (T,ρ). Since we know the true value of ρ, we know whether each test leads to a good decision, ie not rejecting when ρ is zero and rejecting when it is not zero. In Table 1, the results in bold correspond to wrong decisions when ρ differs from zero; when ρ is equal to zero, the decision is correct in all cases. Each test is done at the level of posterior confidence equal to 0.95. Table 2 gives detailed results on the posterior mean and standard deviation of ρ in (5.3), and Table 3 gives the same results on ρ and δ in (5.4). For example, for $T = 15$, one sees in Table 2 that $E(\rho|y)/\sigma(\rho|y)$ is in absolute value less than 2.16 (the 0.975 quantile of the t distribution with 13 degrees of freedom), except for ρ equal to -0.9 or -0.99; hence the statement in Table 1 that BRT "accepts" H_0: $\rho = 0$ but for $\rho = -0.9$ or -0.99. In Table 3, for $T = 100$, one sees that for positive ρ, $E(\delta|y)/\sigma(\delta|y)$ is in absolute value larger than 1.98 (the 0.975 quantile of the t distribution with 96 degrees of freedom), hence the first step of the CFS test rejects $\delta = 0$, a wrong decision. Nevertheless, one sees in Table 4 that CFJT rejects correctly the joint hypothesis ($\delta = 0$ and $\rho = 0$), except for $\rho = -0.2$.

Table 1
Comparison of tests for a single sample when $\rho \neq 0$

	T=15	T=30	T=60	T=100
CFST	**H. accepted** but **$\rho=-0.9$ or ±0.99**	**H. accepted** but **$\rho=\pm0.9$ or ±0.99**	Good decisions but **$\rho=\pm0.2$**	Good decisions but **$\rho\geq-0.2$**
CFJT	**H. accepted** but **$\rho=-0.9$ or -0.99**	**H. accepted** but **$\rho=\pm0.9$ or ±0.99**	Good decisions but **$\rho=\pm0.2$**	Good decisions but **$\rho=-0.2$**
BRT	**H. accepted** but **$\rho=-0.9$ or -0.99**	**H. accepted** but **$\rho=\pm0.9$ or ±0.99**	Good decisions but **$\rho=\pm0.2$**	Good decisions but **$\rho=-0.2$**

CFST: Common Factor Sequential Test (see Section 4)
CFJT: Common Factor Joint Test (see Section 4)
BRT: Bayesian Residuals Test (see Section 3)
Results in bold correspond to wrong decisions.

Table 2
Posterior mean and standard deviation of ρ in augmented regression (5.3)

T \ ρ		-0.99	-0.9	-0.5	-0.2	0.0	+0.2	+0.5	+0.9	+0.99
15	ρ_c	-0.694	-0.554	-0.304	-0.171	-0.072	0.037	0.211	0.630	0.671
	E(ρ|y)	-0.806	-0.580	-0.284	-0.205	-0.130	-0.027	0.152	0.380	0.500
	σ(ρ|y)	0.243	0.288	0.331	0.328	0.323	0.319	0.315	0.290	0.263
30	ρ_c	-0.875	-0.520	-0.227	-0.093	0.021	0.158	0.390	0.743	0.774
	E(ρ|y)	-0.881	-0.582	-0.248	-0.121	-0.010	0.127	0.372	0.668	0.747
	σ(ρ|y)	0.094	0.185	0.223	0.224	0.221	0.216	0.198	0.150	0.132
60	ρ_c	-0.967	-0.842	-0.468	-0.215	-0.051	0.114	0.380	0.824	0.968
	E(ρ|y)	-0.988	-0.845	-0.491	-0.251	-0.093	0.069	0.337	0.760	0.879
	σ(ρ|y)	0.019	0.068	0.120	0.135	0.140	0.142	0.135	0.098	0.088
100	ρ_c	-0.913	-0.844	-0.400	-0.071	0.138	0.341	0.632	0.928	0.985
	E(ρ|y)	-0.926	-0.854	-0.420	-0.089	0.124	0.331	0.630	0.939	0.971
	σ(ρ|y)	0.044	0.054	0.093	0.103	0.103	0.097	0.081	0.039	0.024

E(.|y): posterior mean; σ(.|y): posterior standard deviation
ρ = theoretical autocorrelation (ie population value)
ρ_c = first-order autocorrelation coefficient of generated u_t series

Some points can be noticed about the results of Tables 1 to 4.
i) When the sample size increases, the results of the tests become better since wrong decisions (accepting $\rho = 0$ wrongly) occur progressively for smaller $|\rho|$. For T = 100, good decisions are taken except at $\rho = -0.2$ (with BRT and CFJT).
ii) For all configurations, the decisions of BRT and CFJT are the same; they differ from the decisions of CFST for (T = 100, $\rho \geq 0.2$) and for (T = 15, $\rho = +0.99$).
iii) The posterior mean of ρ in the augmented regressions (5.3) and (5.4) underestimate often ρ by a large quantity when T = 15 or 30 (a well known bias problem). When T is great, this bias tends to vanish. However the posterior means are generally closer to the sampling autocorrelation ρ_c than to the population value of ρ.
iv) There is almost no difference between the posterior means and standard deviations of ρ in the two types of augmented regressions (compare Tables 2 and 3). In Figure 1, one sees that for (T = 100, $\rho = 0$), the posterior Student densities of ρ in (5.3) and (5.4) are almost identical. This is a typical result.
v) In Figure 2, we show the posterior density of PF(ϕ) for the same (T,ρ) configuration as for Figure 1. The density has the same shape as a Fisher density with two degrees of freedom.

Table 3
Posterior mean and standard deviation of ρ and δ in augmented regression (5.4)

T		ρ = -0.99	-0.9	-0.5	-0.2	0.0	+0.2	+0.5	+0.9	+0.99
15	ρ_c	-0.694	-0.554	-0.304	-0.171	-0.072	0.037	0.211	0.630	0.671
	$E(\delta\mid y)$	-3.168	-3.257	-2.772	-2.294	-2.308	-2.466	-2.554	-1.604	-0.902
	$\sigma(\delta\mid y)$	3.022	2.586	2.430	2.446	2.354	2.239	2.083	1.931	2.044
	$E(\rho\mid y)$	-0.780	-0.548	-0.237	-0.174	-0.108	-0.016	0.147	0.363	0.496
	$\sigma(\rho\mid y)$	0.220	0.261	0.313	0.321	0.315	0.309	0.302	0.296	0.277
30	ρ_c	-0.875	-0.520	-0.227	-0.093	0.021	0.158	0.390	0.743	0.774
	$E(\delta\mid y)$	-1.722	-2.324	-2.028	-1.650	-1.558	-1.567	-1.612	-1.016	-0.596
	$\sigma(\delta\mid y)$	2.119	1.744	1.543	1.511	1.463	1.419	1.393	1.435	1.494
	$E(\rho\mid y)$	-0.882	-0.565	-0.218	-0.098	0.008	0.141	0.376	0.664	0.752
	$\sigma(\rho\mid y)$	0.093	0.176	0.246	0.220	0.218	0.211	0.192	0.152	0.132
60	ρ_c	-0.967	-0.842	-0.468	-0.215	-0.051	0.114	0.380	0.824	0.968
	$E(\delta\mid y)$	0.107	0.067	0.063	0.088	0.050	-0.032	-0.180	-0.140	-0.086
	$\sigma(\delta\mid y)$	1.380	1.285	1.101	0.985	0.927	0.891	0.888	0.993	1.032
	$E(\rho\mid y)$	-0.988	-0.846	-0.491	-0.252	-0.095	0.067	0.336	0.759	0.878
	$\sigma(\rho\mid y)$	0.019	0.067	0.119	0.134	0.139	0.141	0.134	0.099	0.086
100	ρ_c	-0.913	-0.844	-0.400	-0.071	0.138	0.341	0.632	0.928	0.985
	$E(\delta\mid y)$	-0.495	-0.257	0.110	0.393	0.590	0.790	1.077	1.455	1.447
	$\sigma(\delta\mid y)$	0.554	0.533	0.437	0.395	0.389	0.310	0.439	0.511	0.521
	$E(\rho\mid y)$	-0.923	-0.855	-0.422	-0.091	0.122	0.331	0.631	0.939	0.970
	$\sigma(\rho\mid y)$	0.040	0.053	0.093	0.101	0.101	0.095	0.078	0.037	0.024

See legend of Table 2.

vi) The numbers N and M (see Section 4) of drawings of the Student posterior (4.4) were set to 1000 in most experiments. A comparison with 10000 draws, needed to obtain smooth graphs of densities, revealed that the other results were accurate enough with 1000 points.

Since the BR and CFJ tests seem to give better results than CFST, we have repeated 1000 sampling experiments to approximate the power of these tests. In Tables 5 to 8, we report the percentages of rejection of BRT and CFJT, and the percentages of acceptance (A), rejection (C) or uncertainty (U) for the classical Durbin-Watson test

Table 4

CFJT: values of PF(0) and of Probability that PF(ϕ) is greater than PF(0)

T \ P	-0.99	-0.9	-0.5	-0.2	0.0	+0.2	+0.5	+0.9	+0.99
15	9.54 0	4.24 0.04	1.31 0.3	0.77 **0.47**	0.66 0.52	0.74 **0.47**	1.14 0.35	1.23 0.32	1.96 **0.18**
30	48.51 0	7.24 0	1.68 0.09	0.79 0.46	0.61 0.54	0.96 0.39	3.06 **0.07**	10.38 0	18.36 0
60	1265.3 0	80.69 0	9.03 0	1.85 **0.18**	0.02 1.0	0.12 **0.9**	3.39 0.04	30.45 0	51.39 0
100	285.64 0	133.69 0	10.56 0	0.90 **0.4**	1.90 0.07	7.83 0	35.46 0	322.46 0	801.63 0

First line: PF(0); second line: P [PF(ϕ)≥PF(0)|y]

Table 5
Power Function (T=15)

	P	-0.99	-0.9	-0.5	-0.2	0.0	+0.2	+0.5	+0.9	+0.99
	A	1.1	3.5	32.9	69.2	79.9	78.4	49.5	15.4	9.4
DW:	C	97.8	91.6	43.4	11.6	5.9	6.0	26.9	69.3	69.7
	U	1.1	4.9	23.7	19.2	14.2	15.6	23.6	15.3	20.9
BRT		98.7	94.2	43.4	9.2	2.9	2.4	12.9	60.3	67.8
CFJT		98.9	91.8	34.0	8.1	3.9	4.0	17.8	62.2	70.6

Table 6
Power Function (T=30)

	P	-0.99	-0.9	-0.5	-0.2	0.0	+0.2	+0.5	+0.9	+0.99
	A	0	0.1	11.5	61.3	83.3	69.4	19.5	0.5	0.2
DW:	C	100	99.7	80.2	23.3	8.0	16.5	68.3	98.7	98.9
	U	0	0.2	8.3	15.1	8.7	14.1	12.2	0.8	0.9
BRT		100	99.7	77.3	18.7	3.6	9.8	60.2	98.9	99.3
CFJT		100	99.7	67.4	15.3	5.5	8.8	51.2	97.5	98.7

Table 7
Power Function (T=60)

	P	-0.99	-0.9	-0.5	-0.2	0.0	+0.2	+0.5	+0.9	+0.99
	A	0	0	0.9	48.9	89.2	53.9	2.1	0	0
DW:	C	100	100	98.2	40.3	6.1	34.8	96.9	100	100
	U	0	0	0.9	10.8	4.7	11.3	1.0	0	0
BRT		100	100	97.9	37.1	5.1	24.5	94.8	100	100
CFJT		100	100	95.7	28.2	5.1	18.2	91.9	100	100

Table 8
Power Function (T=100)

ρ		-0.99	-0.9	-0.5	-0.2	0.0	+0.2	+0.5	+0.9	+0.99
	A	0	0	0	33.1	89.0	34.8	0.1	0	0
DW:	C	100	100	100	58.6	7.5	57.4	99.9	100	100
	U	0	0	0	8.3	3.5	7.8	0	0	0
BRT		100	100	100	53.4	3.4	44.4	99.2	100	100
CFJT		100	100	99.3	42.2	4.2	33.4	99.3	100	100

(with nominal size 10%). For example, if $\rho = 0$ and $T = 15$, one rejects H_0: $\rho = 0$ in 3.9 % of the cases with CFJT, in 2.9% with BRT, in 5.9% with DW (at least since there are 14.2% cases of "uncertainty").

The results show that the power of all tests is sizable as soon as $|\rho|$ exceeds 0.2 even with a small sample size. The obtained powers cannot be compared directly since the sizes differ. For T larger than 60, BRT is more powerful than CFJT. A possible reason is that BRT imposes the common factor restriction, and that the alternative hypothesis of CFJT is too wide (as explained in Section 4), thus requiring a test of two restrictions, instead of one as BRT. Since BRT is easy to compute and is available in BIP, *it becomes our recommended test*, at least for a test of first-order autocorrelation. We notice finally that BRT is invariant to the value of the variance of the innovation ε_t assumed in our Monte-carlo setup, while this is not the case of the common factor tests.

Nevertheless, the PF test seems to be a useful procedure to characterise approximately a HPD region in a large dimension (as does the F-test exactly for linear restrictions) and to solve the problem introduced by a non-linear function of the coefficients. In particular the first step of CFST may be of interest in itself in a search of a parsimonious parameterisation of an ADL model.

6. Conclusion

Despite the conclusions emerging from the Monte-Carlo experiments reported in the previous section, further research is needed to support more fully these conclusions. In particular, sensitivity analyses with respect to the value of the regression coefficients and with respect to the process of the exogenous variable are needed. Simulations when the prior is informative or for higher order autocorrelation processes (such as p = 4) are also on the agenda. Another topic for further research is the use of the approximate posterior densities of α from the augmented regressions as importance functions for the exact posterior density, and the comparison of importance sampling with Gibbs sampling for inference on α.

Graph 1: Posterior densities of ρ, T=100, ρ_0=0.0

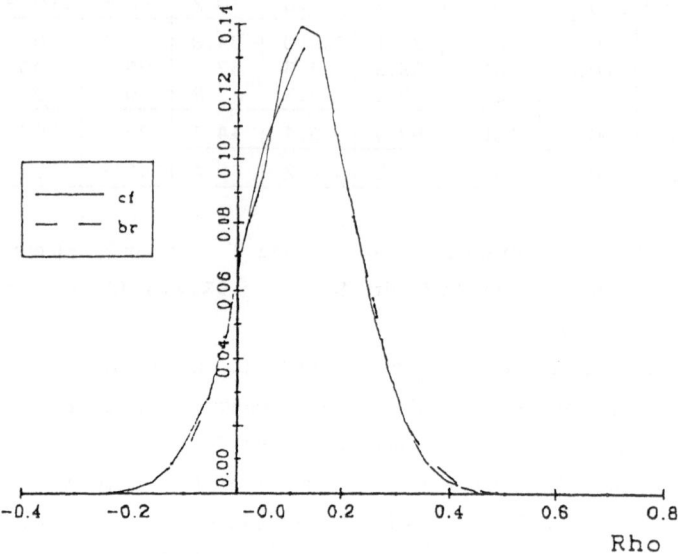

Graph 2: Posterior density of PF, T=100, ρ_0=0.0

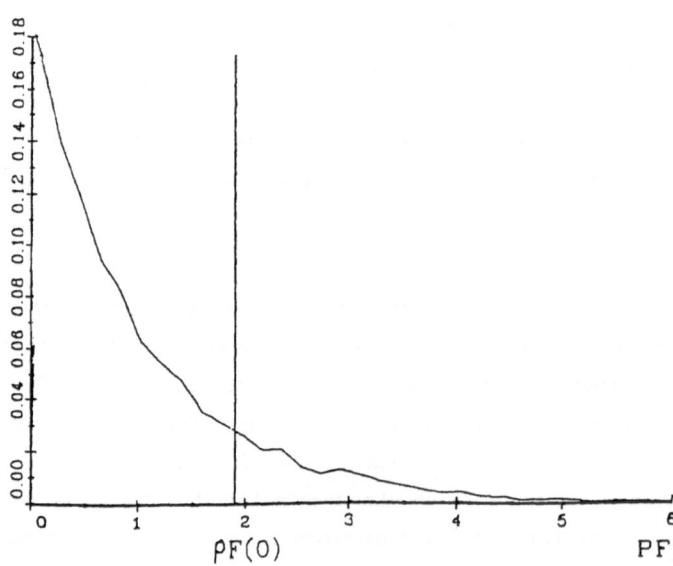

Acknowledgment

We are indebted to Michel Mouchart and to Jean-Marie Rolin for their comments on the previous version of this paper. The usual disclaimer applies.

References

Aprahamian, F., Bauwens, L., and Lubrano, M. (1990), A Presentation of BIP. Manuscript, GREQE, Marseille.

Aprahamian, F., Lubrano, M. and Marimoutou, V. (1990), A Bayesian Approach to Misspecification Tests: the U.K. Money Demand Equation Revisited. GREQE DP 90A07, Marseille.

Bauwens, L. and Lubrano, M. (1991), Bayesian Diagnostics for Heterogeneity. *Annales d'Economie et de Statistique*, 20/21, 17-40.

Bauwens, L. and Lubrano, M. (1992), Trends and Break Points in Bayesian Econometrics. Paper presented at 'Economics: the Next Ten Years', Conference for the tenth anniversary of GREQE, manuscript, CORE, Louvain-La-Neuve.

Drèze, J.H. and Mouchart, M. (1990), Tales of Testing Bayesians. In *Contributions to Econometric Theory and Application. Essays in Honour of A.L. Nagar*, R.A.L. Carter, Dutta, J. and Ullah, A. (eds). Springer-Verlag, New-York.

Florens, J.-P. and Mouchart, M. (1989), Bayesian Specification Tests. In *Contributions to Operations Research and Economics*, B. Cornet and H. Tulkens (eds). Cambridge University Press, Cambridge.

Harvey, A.C. (1990), *The Econometric Analysis of Time Series* (second edition). Philip Allan.

Hausman, J.A. (1978), Specification Tests in Econometrics. *Econometrica*, 46, 1251-1271.

Hendry, D.F. and Mizon, G.E. (1978), Serial Correlation as a Convenient Simplification, not a Nuisance: a Comment on a Study of the Demand for Money by the Bank of England. *Economic Journal*, 88, 549-563.

Pagan, A. (1978), Detecting Autocorrelation After Bayesian Regression. CORE DP 7825, Louvain-La-Neuve.

Poirier, D.J. (1987), Bayesian Diagnostic Testing in the General Linear Normal Regression Model. Paper presented at the third Valencia Meeting on Bayesian Statistics.

Richard, J-F. (1977), "Bayesian Analysis of the Regression Model when the Disturbances are Generated by an Autoregressive Process. In *New Developments in the Application of Bayesian Methods*, A. Aykac and C. Brumat (eds). North-Holland Publishing Cy, Amsterdam.

Zellner, A. (1971), *An Introduction to Bayesian Inference in Econometrics*. John Wiley, New York.

Intensive Numerical and Symbolic Computing in Parametric Test Theory

Dirk Wauters and Lea Vermeire
Katholieke Universiteit Leuven Campus Kortrijk
Universitaire Campus, B-8500 Kortrijk, Belgium

Key words : gamma family, geodesic test, locally most mean power test, Weibull family.

Abstract

In the construction of multiparameter significance tests intensive computing may be involved at two levels : *numerical* computing in the construction of critical regions with particular optimality properties, *symbolic* computing to obtain a better understanding of a statistical model e.g. by giving the geometry of the model. Each level is explained in the construction of two tests for a simple null hypothesis in two lifetime models.

The g-LMMPUα test maximizes the local mean power with respect to a metric g under all unbiased tests. This test is obtained for the two-parameter gamma family. The construction of the critical region involves the solution of a system of non-linear equations, two-dimensional numerical quadrature, numerical Fourier inversion and interpolation. The NAG Fortran library is a basic tool.

The *geodesic* test uses the Rao distance (information distance) between the ML-estimators and the null hypothesis as test statistic. This test is obtained for the two-parameter Weibull family. The symbolic computation of tensor components and connection symbols requires essentially the partial derivatives of the loglikelihood and expectations. The Mathematica language allows direct symbolic computation of partial derivatives. For the expectations a *package* within Mathematica is constructed.

1 LMMPU-test for the gamma family.

1.1 Definition and sufficient statistics.

A gamma random variable X is a univariate positive random variable with density function

$$p(x; \alpha, \beta) = \frac{1}{\Gamma(\alpha)\beta^\alpha} x^{\alpha-1} e^{-x/\beta} \quad x \geq 0, \ \alpha, \beta > 0,$$

where α is a shape parameter and β is a scale parameter. It is an exponential family with natural parameters $(\alpha, 1/\beta)$ and natural sufficient statistic $(\ln(X), -X)$ based on one observation X.

Hence an i.i.d. sample of size n summarizes in the sufficient statistics

$$
\begin{cases}
S = \dfrac{1}{n} \displaystyle\sum_{i=1}^{n} X_i = \bar{X} \\[2mm]
P = \dfrac{1}{n} \displaystyle\sum_{i=1}^{n} \ln(X_i) = \ln\left(\left(\prod_{i=1}^{n} X_i\right)^{1/n} \right) \;,
\end{cases}
\tag{1}
$$

or the pair $(\bar{X}, \tilde{X})$ of the arithmetic mean and the geometric mean of the sample.

The statistic

$$
U_n = \frac{e^P}{S} = \frac{(\prod_{i=1}^{n} X_i)^{1/n}}{\sum_{i=1}^{n} X_i/n} = \frac{\tilde{X}}{\bar{X}}
\tag{2}
$$

with range $[0,1]$ and its negative logarithm

$$
Y = -\ln(U_n) = \ln(S) - P
\tag{3}
$$

are frequently used. The parametric structure of the joint (S, P)-density is given by the factorization

$$
\mathrm{P}_{(S,P)}(s, p; \alpha, \beta) = \frac{e^{n\alpha p} e^{-ns/\beta}}{[\Gamma(\alpha)]^n \beta^{n\alpha}} \, \mathrm{f}_n(s, p), \quad s \geq e^p
\tag{4}
$$

where f_n is independent of α, β ; indeed

$$
\mathrm{f}_n(s, p) = \frac{n^n}{\Gamma(n)} s^{n-2} e^{(1-n)p} \mathrm{g}_n\left(\frac{e^p}{s}\right)
\tag{5}
$$

where g_n is the density of $U|_{\alpha=1}$.

Using the loglikelihood of the gamma family,

$$
\ell(x; \alpha, \beta) = \ln \mathrm{p}(x; \alpha, \beta) = -\ln\Gamma(\alpha) - \alpha \ln(\beta) + (\alpha - 1)\ln(x) - x/\beta,
$$

the information matrix (g_{ij}), where $g_{ij} = \mathrm{E}\{\partial_i \ell(x; \theta)\partial_j \ell(x; \theta)\} = -\mathrm{E}\{\partial_i \partial_j \ell(x; \theta)\}$ and $\theta = (\alpha, \beta)$, and its inverse are obtained :

$$
(g_{ij}) = \begin{pmatrix} \psi'(\alpha) & \dfrac{1}{\beta} \\[2mm] \dfrac{1}{\beta} & \dfrac{\alpha}{\beta^2} \end{pmatrix}, \qquad (g^{ij}) = \frac{1}{\alpha\psi'(\alpha) - 1} \begin{pmatrix} \alpha & -\beta \\ -\beta & \beta^2 \psi'(\alpha) \end{pmatrix}.
$$

1.2 The g-LMMPU critical region.

Consider the simple null hypothesis $H_0 : (\alpha, \beta) = (\alpha_0, \beta_0)$ on the two parameters simultaneously. The *locally most mean power and unbiased level $\underline{\alpha}$ test with respect to a metric g* or g-LMMPU$\underline{\alpha}$ test (Sen Gupta and Vermeire, 1982, 1986, Van Lindt, 1984, Vermeire and Wauters, 1988) maximizes the mean power on a g-spherical neighbourhood of the null hypothesis. For the gamma family this test results in the critical region w :

$$
g^{\alpha\alpha}(P - b)^2 + \frac{2}{\beta^2}g^{\alpha\beta}(P - b)(S - a) + \frac{1}{\beta^4}g^{\beta\beta}(S - a)^2 \geq c^2,
\tag{6}
$$

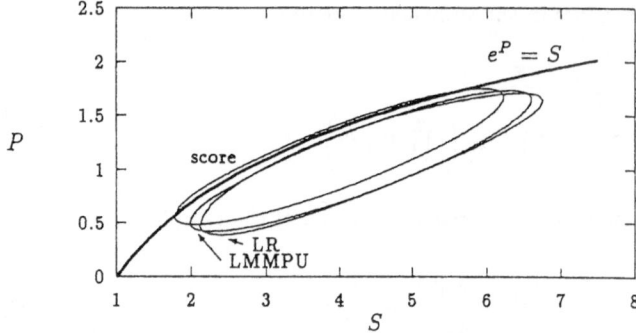

Figure 1: Critical region for LMMPU-, LR- and score test, $n = 10$, $\underline{\alpha} = 0.05$.

where g^{ij}, β all must be evaluated at the null hypothesis (α_0, β_0), and where the constants a, b, c have to be determined from the LU$\underline{\alpha}$ conditions :

$$\int_{\bar{w}} p_0(s,p)dsdp = 1 - \underline{\alpha} \tag{7}$$

$$\int_{\bar{w}} s\, p_0(s,p)dsdp = \alpha_0\beta_0(1 - \underline{\alpha}) \tag{8}$$

$$\int_{\bar{w}} p\, p_0(s,p)dsdp = (\ln(\beta_0) + \psi(\alpha_0))(1 - \underline{\alpha}) \tag{9}$$

where $\bar{w} = \mathbb{R}^2 \setminus w$ is the acceptance region and $p_0(s,p)$ is the joint density of the sufficient statistic (S, P) under H_0. These conditions are derived from the conditions $B(\alpha_0, \beta_0) = \underline{\alpha}$, $D_\alpha B(\alpha_0, \beta_0) = 0$, $D_\beta B(\alpha_0, \beta_0) = 0$ on the power function B, and the structure formula (4). The critical region w is the outer region of an ellipse in the (S, P)-plane. Shape and direction can easily be obtained.

1.3 Computational aspects.

To obtain the constants a, b, c in (6) from (7)–(9), subject to a specified accuracy, three problems should be solved : the solution of a system of non-linear equations, numerical two-dimensional quadrature and a good approximation to the U_n-density. These constants depend on the null hypothesis (α_0, β_0) and the metric in the null hypothesis g_0, the sample size n and the significance level $\underline{\alpha}$.

1.3.1 The system of non-linear equations.

The NAG fortran routine C05NCF was used to solve (7)–(9). The NAG library provides two routines for the iterative solution - by a modification of the Powell hybrid method - of a system

$$\begin{cases} f_1(x_1,\ldots,x_n) &= 0 \\ \quad\cdots \\ f_n(x_1,\ldots,x_n) &= 0 \end{cases}$$

where $f_1,\ldots,f_n$ are $\mathbb{R}^n \mapsto \mathbb{R}$ functions. Both construct a next approximation to the solution from a given approximation (e.g. the starting value) and the Jacobian matrix of the system. The routines differ in the way they compute the Jacobian matrix, that is the partial derivatives of the functions f_i : in the routine C05PCF the user enters the partial derivative functions as a computational rule, whereas in the routine C05NCF the partial derivatives are approximated by forward differences of the functions f_i. For the system (7)–(9) under consideration the first routine would require 9 additional double integrals in each iteration. Therefore the routine C05NCF is used. Important parameters are the starting values, the parameter EPSFCN which specifies the order of largest relative error in the functions f_i and hence measures the accuracy of the integrals, the parameter XTOL which is the tolerated relative error (or the required accuracy) in the solution. The initial values are extremely important for the solution. They can be chosen as the asymptotic values $(a,b,c) = (\alpha_0\beta_0, \psi(\alpha_0) - \ln(\beta_0), \chi^2_{2,1-\alpha})$ or as the solution under a previous parameter n. As the convergence test of C05NCF is based on the relative error the unknowns should be scaled to equal magnitude. The accuracy of the integrals should be at least a factor 10 higher than the accuracy parameter XTOL.

1.3.2 The double integrals.

The NAG routine D01DAF is used for the evaluation of the double integrals. Each integral has the form

$$I = \int_{y_a}^{y_b} \int_{\varphi_1(y)}^{\varphi_2(y)} f(x,y)dxdy$$

and is evaluated by two univariate integrals

$$I = \int_{y_a}^{y_b} F(y)dy, \quad F(y) = \int_{\varphi_1(y)}^{\varphi_2(y)} f(x,y)dx,$$

which are called the outer and the inner integral respectively. The functions $\varphi_1(y)$ and $\varphi_2(y)$ and the limits y_a, y_b are easily computed from the equation of the ellipse (6). For the univariate integrals, the subroutine D01DAF uses the method of Patterson, which uses growing interlacing sets of interpolation points.

The routine may overlook narrow peaks and is limited to 255 interpolating points on the interval of integration, which may lead to convergence errors. Therefore two modifications are made on the routine. First, the region is divided in 4 subregions if necessary and the results are added up. Secondly, the source code of the library routine was modified to use at least 31 points for the outer integral and at least 15 points for the inner integral.

1.3.3 The U_n-density.

As to the computation of the U_n-density much research has been done on numerical approximations : Bain & Engelhardt (1975), Glaser (1976), Nandi (1980), Jensen (1986), Bowman & Shenton (1988). For the practical applications in this paper we computed the U_n-density as a mixture of the approximation by Glaser at the upper part and a cubic spline approximation at the lower tail of the density, in the case $n \leq 15$. For large n the chisquare approximation of Bain & Engelhardt (1975) was used.

Glaser (1976) obtained the following formula for the density of U_n

$$g_{\alpha,n}(u) = \left[\prod_{r=1}^{n-1} \frac{\Gamma(\alpha + \frac{r}{n})}{\Gamma(\alpha)} \right] \frac{n^{(n-1)/2}}{\Gamma((n-1)/2)} u^{n\alpha-1}(-\ln(u))^{\frac{n-3}{2}} \xi_n(u^n)$$

where $\xi_n(x)$ is the series

$$\xi_n(u^n) = \sum_{j=0}^{\infty} n^j \nu_j(n)(-\ln(u))^j, \quad e^{-2\pi/n} \leq u \leq 1, \tag{10}$$

$$\nu_j(n) = \sum_{J} \frac{\Gamma((n-1)/2)}{\Gamma((n-1)/2+j)} \frac{y_1^{s_1} y_3^{s_3} \cdots y_d^{s_d}}{s_1! s_3! \cdots s_d!}, \tag{11}$$

d denotes the largest odd integer $\leq n$,

$$J = \{(s_1, s_3, \ldots, s_d)| s_k \in \mathbb{N}, d \text{ odd}, 1 \cdot s_1 + 3 \cdot s_3 + \cdots + d \cdot s_d = j\},$$

$$y_k = \frac{(-1)^{k+1}}{k(k+1)}(n - n^{-k}) B_{k+1}$$

and B_k are the Bernoulli numbers. A pascal program was written to compute the possible combinations for the distinct terms (11) and to generate a Mathematica program which produces the functions $\nu_j(n)$, as symbolic expressions in n, for $j = 1, \ldots, 35$.

The u-interval is imposed by the series convergence condition. Numerical computations show that for $n \leq 12$ this interval absorbs at least half the probability mass of U_n.

The density in the lower tail is approximated by spline interpolation on a set of discrete points, at which the density should be available. To construct the cubic spline NAG routine E02BAF is used, to evaluate the spline E02BBF is used.

The density in the interpolation points is obtained by numerical Fourier inversion. Let $\alpha = 1$. The Laplace transform of the Y-density $\text{p}_n(y)$ is

$$F(z) = \mathcal{L}\{\text{p}_n(y)\} = \text{E}(e^{-Yz}) = \frac{n^z \Gamma(n) \Gamma^n(1 + \frac{z}{n})}{\Gamma(n + z)}$$

and, by inversion

$$\text{p}_n(y) = \frac{1}{2\pi i} \int_{\sigma - i\infty}^{\sigma + i\infty} e^{yz} F(z) dz, \qquad Re(\sigma) > -n. \qquad (12)$$

The integrand has poles of order $(n - 1)$ at $z = -n, -2n, -3n, \ldots$ A straightforward solution would be to integrate (12) numerically, i.e. to integrate the imaginary part of the integrand. This function is strongly oscillating and very difficult to integrate. Therefore Talbot's method for the Fourier inverse is used, in a slightly modified version which improves on the speed of convergence. Talbot proposes to replace (12) by

$$\text{p}_n(y) = \frac{\lambda}{2\pi i} \int_C e^{\lambda z + \sigma} F(\lambda z + \sigma) dz,$$

$$C = \{z \in \mathbb{C} | z = \theta \cot(\theta) + i\mu\theta, -\pi < \theta < \pi\}$$

which reduces to

$$\frac{(ne^y)^\sigma \lambda \Gamma(n)}{\pi} \int_0^\pi \Im \left[\exp\left\{ \lambda(\theta \cot(\theta) + i\mu\theta)(y + \ln(n)) + n \ln \Gamma(1 + \frac{\lambda}{n}(\theta \cot(\theta) + i\mu\theta) + \frac{\sigma}{n}) \right. \right.$$

$$\left. \left. - \ln \Gamma(n + \lambda(\theta \cot(\theta) + i\mu\theta) + \sigma) \right\} \left(\frac{1}{\tan(\theta)} - \frac{\theta}{\sin^2(\theta)} + i\mu \right) \right] d\theta.$$

The integrand is finite at $\theta = 0$ and tends to zero as $\theta \to \pm\pi$. The parameters λ and σ are chosen such that the contour C encloses the singularities of $F(\lambda z + \sigma)$. Our choice is $\lambda = 1$, $\sigma = -n + 1$ and $\mu = 1 + nu$. The integral is calculated with the NAG library routine D01AHF, which integrates definite integrals over a finite range with an adaptive method. This method gives excellent convergence and a relative accuracy of 10^{-6}. The original Talbot method used $\mu = 1$. The above method gives more flexibility to adjust the contour to the parameter n in the integrand.

It is an essential part in the above calculations to obtain a very good approximation for the logarithm of the complex gamma function. A routine is available from the authors. This routine was inspired by the FORTRAN subroutine for the polygamma functions in Bowman & Shenton (1988) pp 247–248. It relies on three formulas : Stirling's approximation

$$\ln \Gamma(z + 1) = (z + \frac{1}{2}) \ln(z) - z + \frac{1}{2} \ln(2\pi) + \frac{B_2}{2z} + \frac{B_4}{3.4z^3} + \frac{B_6}{5.6z^5} + \cdots,$$

the reflexion formula

$$\Gamma(z) = \frac{(-1)^m \pi}{\Gamma(1 - z) \sin(\pi(z + m))}, \quad m = 0, 1, 2, \ldots,$$

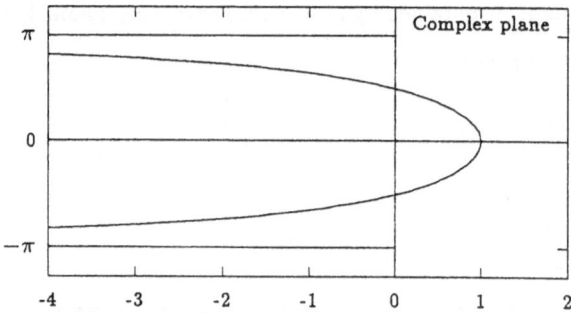

Figure 2: The Talbot contour.

and the recurrence formula

$$\Gamma(z+1) = z\Gamma(z).$$

The Bernoulli numbers are obtained from Mathematica or from adequate tables.

2   A geodesic test for the Weibull family.

2.1   Definition.

The two-parameter Weibull distribution $W(b,c)$ has the density

$$p(x;b,c) = \frac{c}{b}\left(\frac{x}{b}\right)^{c-1} e^{-\left(\frac{x}{b}\right)^c}, \quad x > 0, \tag{13}$$

where $b > 0$, $c > 0$. The cumulative distribution function is

$$F(x;b,c) = 1 - \exp\left(-\left(\frac{x}{b}\right)^c\right), \quad x > 0.$$

The parameter b is called the *scale* parameter and c is the *shape* parameter.

2.2   Geometric structure of the Weibull manifold : the information matrix, the alpha-connections and the alpha-curvature.

The geometric structure of a family of distributions, as derived in the book by Amari e.a. (1987) consists of the information metric, the α-connections, the skewness tensor and the curvatures. They can be computed in symbolic form with the aid of the Mathematica system. Indeed, computationally these tensor components and connection symbols on a statistical manifold require essentially the expected partial derivatives of the loglikelihood and their partial derivatives :

$$\ell(x;\theta) \quad = \quad \ln p(x;\theta) \text{ , the log-likelihood function}$$

$$(g_{ij}) \quad = \quad -\mathrm{E}\{\partial_i\partial_j\ell(x;\theta)\} \text{ , the information metric}$$

$$\Gamma_{ijk} \quad = \quad \tfrac{1}{2}\left[\partial_i g_{jk} + \partial_j g_{ik} - \partial_k g_{ij}\right] \text{ , the Christoffel symbols of the first kind}$$

$$\Gamma^k_{ij} \quad = \quad g^{k\ell}\Gamma_{ij\ell} \text{ , the Christoffel symbols of the second kind}$$

$$R_{ijkm} \quad = \quad [\partial_i\Gamma^s_{jk} - \partial_j\Gamma^s_{ik}]g_{sm} + [\Gamma_{irm}\Gamma^r_{jk} - \Gamma_{jrm}\Gamma^r_{ik}] \text{ , the curvature tensor}$$

$$K \quad = \quad \tfrac{1}{2}R_{ijkm}g^{im}g^{jk} = \tfrac{R_{1221}}{\det(g)} \text{ , the Gaussian curvature}$$

$$T_{ijk} \quad = \quad \mathrm{E}\{\partial_i\ell\partial_j\ell\partial_k\ell\} \text{ , the skewness tensor}$$

$$\overset{(\alpha)}{\Gamma}_{ijk} \quad = \quad \Gamma_{ijk} - \tfrac{\alpha}{2}T_{ijk} \text{ , the } \alpha\text{-connection Christoffel symbols}$$

$$\overset{(\alpha)}{R}_{ijkm} \quad \quad \text{ , the } \alpha\text{-connection curvature tensor}$$

$$\overset{(\alpha)}{K}(p) \quad = \quad \tfrac{1}{2}\overset{(\alpha)}{R}_{ijkm}\,g^{im}g^{jk}$$

$$\quad = \quad \tfrac{1}{2}\frac{\overset{(\alpha)}{R}_{1221} - \overset{(\alpha)}{R}_{1212}}{\det(g)} \text{ , the } \alpha\text{-Gaussian (sectional) curvature.}$$

The Mathematica language allows direct symbolic computation of partial derivatives and expectations (integrals). A particular statistical model leads to specific integrals. If Mathematica does not recognize these integrals there is no output. Often this problem can be solved by defining a *package* within Mathematica for the expectations : the expectation operator is linear, hence the expected partial derivatives reduce to sums of moments; often these moments can be computed recursively.

This happened with the Weibull family, under consideration here. The partial derivatives were obtained. For the expectations a package was constructed based on

1. the linearity of the expectation operator E :

$$\mathrm{E}(aX + bY) = a\mathrm{E}(X) + b\mathrm{E}(Y) \quad ;$$

2. recursive computations of the moments $\mathrm{E}((X^c)^k \ln(X)^n)$ by

$$\mathrm{E}((X^c)^k \ln(X)^n) = \frac{b^{kc}}{c^n}\mathrm{E}(Y^k \ln(Y)^n) - \mathrm{E}\left(X^{kc}(\ln(X) - \ln(b))^n - \ln(X)^n)\right)$$

where

$$\mathrm{E}(Y^k \ln(Y)^n) = \left.\frac{d^n\Gamma(u)}{du^n}\right|_{u=k+1}$$

The information metric tensor $i(b,c)$ is given by its matrix

$$(g_{ij}) = \begin{pmatrix} \dfrac{c^2}{b^2} & \dfrac{\gamma - 1}{b} \\[2mm] \dfrac{\gamma - 1}{b} & \dfrac{1}{c^2}[\dfrac{\pi^2}{6} + (1 - \gamma)^2] \end{pmatrix}$$

and the curvatures are

$$K(p) = \frac{R_{1221}}{\det(g)} = -\frac{6}{\pi^2},$$

$$\overset{(\alpha)}{K}(p) = -\frac{6}{\pi^2} + \frac{18\alpha^2}{\pi^4}(\pi^2 - 2 - 4\zeta(3)),$$

where ζ is the Riemann zeta function.

2.3   The information distance.

In this subsection an explicit formula for the distance between two Weibull distributions, under the Rao information metric is obtained. Under the information geometry ($\alpha = 0$, the Riemannian geometry) the Gaussian curvature is a negative constant $-6/\pi^2$. Hence, the Weibull model is a hyperbolic space or Poincaré half plane and there exists a parameterization which provides an explicit formula for the (Riemannian) distance between two distributions in the model. This in turn will allow exact level α tests in the next section. The information distance between two Weibull distributions $W(b,c)$ and $W(b_0,c_0)$ equals

$$d((b,c),(b_0,c_0)) = \frac{\pi}{\sqrt{6}} \text{Argch} \left\{ \frac{cc_0}{2} \left[\frac{6}{\pi^2} \left(\ln(\frac{b}{b_0}) + (1-\gamma)(\frac{1}{c} - \frac{1}{c_0}) \right)^2 + \left(\frac{1}{c^2} + \frac{1}{c_0^2} \right) \right] \right\}, \quad (14)$$

This result can also be obtained from the distance formula of Yoshizawa (1971) in a location-scale family, and the property that a Weibull-family can be transformed into a location-scale family.

2.4   The i-geodesic test.

Using the ML-statistic $(\hat{b}, \hat{c})$ and the distance formula (14) the geodesic test statistic under H_0 : $(b,c) = (b_0,c_0)$ is

$$T = n\text{d}^2((\hat{b}, \hat{c}), (b_0, c_0)).$$

H_0 is rejected if T becomes too large : $T > d$, where the constant d is determined by the significance level α. For large samples, by the asymptotic behaviour of this test (Burbea & Oller, 1989), a good approximation for d is the $1 - \alpha$ upper quantile of the χ_2^2 distribution.

For small samples this constant has to be determined numerically; as the joint density of the ML-estimators is not available a table with the small sample quantiles of T has been constructed through Monte-Carlo simulations and is available upon request. The intensive computing involved was reduced, as the null distribution of T depends on the sample size only and not on the particular null hypothesis - a result which the authors could prove for general location-scale families (Wauters & Vermeire, 1992), as an application of the pivotal property of the ML-statistic in a location-scale family (Antle & Bain, 1969).

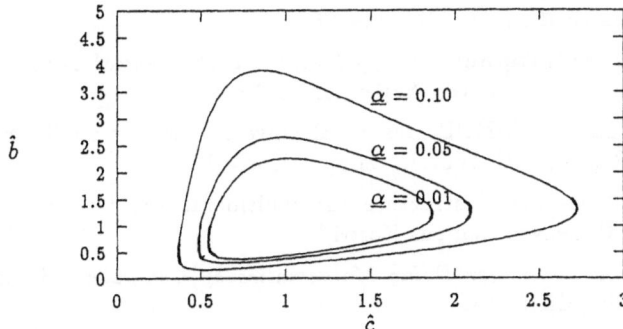

Figure 3: Critical region for the geodesic test.

References

Abramowitz, M. and Stegun, I. A. (1972) *Handbook of Mathematical Functions with Formulas, Graphs and Mathematical Tables.* Wiley, New York.

Amari, S.-I., Barndorff-Nielsen, O. E., Kass, R. E. Lauritzen, S. L., and Rao, C. R. (1987) Differential geometry in statistical inference. *Lecture Notes - Monograph Series 10.* Institute of Mathematical Statistics, Hayward (California).

Antle, C. E. and Bain, L. J. (1969) A property of maximum likelihood estimators of location and scale parameters. *SIAM Review,* 11, 2, 251–253.

Bain, L. J. and Engelhardt, M. (1975) A two-moment chi-square approximation for the statistic $\log(\bar{x}/\tilde{x})$. *Journal of the American Statistical Association,* 70, 352, 948–950.

Bowman, K. O. and Shenton, L. R. (1988) *Properties of Estimators for the Gamma Distribution.* Marcel Dekker, New York.

Burbea, J. and Oller, J. M. (1989) On Rao distance asymptotic distribution. *Mathematical Preprint Series No. 67.* Universitat de Barcelona.

Glaser, R. E. (1976) The ratio of the geometric mean to the arithmetic mean for a random sample from a gamma distribution. *Journal of the American Statistical Association,* 71, 354, 480–487.

Jensen, J. L. (1986) Inference for the mean of a gamma distribution with unkwown shape parameter. *Scandinavian Journal of Statistics,* 13, 135–151.

Nandi, S. B. (1980) On the exact distribution of a normalised ratio of the weighted geometric mean to the unweighted arithmetic mean in samples from gamma distributions. *Journal of the American Statistical Association,* 75, 369, 217–220.

The Numerical Algorithms Group Ltd (1987) *The NAG Fortran Library Manual - Mark 12.* NAG Central Office, Oxford.

Piessens, R., De Doncker-Kapenga, E., Überhuber, C., and Kahaner, D. (1983) *QUADPACK, A Subroutine Package for Automatic Integration.* Springer-Verlag, Berlin.

Sen Gupta, A. and Vermeire, L. (1982) Locally optimal tests for multiparameter hypotheses. Technical Report 671, Department of Statistics, Univ. of Wisconsin, Madison.

Sen Gupta, A. and Vermeire, L. (1986) Locally optimal tests for multiparameter hypotheses.

Journal of the American Statistical Association, 81, 397, 818–825.

Van Lindt, D. (1984) *Statistical Tests for Multiparameter hypotheses in a Differential Geometric Framework.* (In Dutch). Doctoral thesis, K.U. Leuven, Supervisor L. Vermeire.

Vermeire, L. and Wauters, D. (1988) The $g-$LMMPUα test for the inverse Gaussian family with a simple null hypothesis. Technical Report, K.U. Leuven, Campus Kortrijk.

Wauters, D. and Vermeire, L. (1992) Geometric structure and multiparameter tests for the Weibull family. Technical Report, K.U. Leuven Campus Kortrijk.

Wolfram, S. (1991) *Mathematica : A System for Doing Mathematics by Computer.* Second edition. Addison-Wesley, Redwood City, California.

Yoshizawa, T. (1971) A geometrical interpretation of location and scale parameters. Memorandum TYH-3, University of Tokyo, Japan.

Learning Data Analysis and Mathematical Statistics with a Macintosh

Anestis Antoniadis[1], Jacques Berruyer[2] and René Carmona[3]

[1] Lab. IMAG–LMC, Univ. Joseph Fourier, B.P. 53 X, 38041 Grenoble (France).
[2] Département de Mathématiques, Univ. de Saint-Etienne, 42100 Saint-Etienne.
[3] Department of Mathematics, University of California, Irvine, CA 92717, USA.

ABSTRACT

The purpose of this note is to report on pedagogical experiments centered around the use of a dedicated program in the learning of mathematical statistics and the practice of data analysis. The authors developed a set of lectures notes with a strong emphasis on: Graphical analyses, random number generation, simulation techniques, resampling methods, dynamic illustration of regression diagnostics, robust methods . . . . Most of these concepts can be presented to undergraduate students but no appropriate textbook existed. Also, such a pedagogical experiment could not be conceived without the use of a computer program for use as a companion to the lectures. No satisfactory solution could be found from the existing commercial or public softwares. We discuss in detail some of the most salient features of the experiment and we describe the tools which the authors developed in the process.

Key Words: Dynamic graphics, software, teaching of statistics.

1. Introduction

Nowadays, statistical analyses are needed in more branches of sciences and engineering, whether their involvement concerns the design of experiments, of expert systems or the treatment of large data sets. The sizes of the data sets which are amenable to investigations increased dramatically during the last decades due to the advent of new generations of computers. New computational techniques were created to take full advantage of these new capabilities. A new field has been created and the name of *computational statistics* is often appropriately used for it. Unfortunately, the teaching of statistics did not always follow this evolution and it has still too often to catch up with the successes of this new discipline.

Many attempts to bridge this gap have been made. It is very common to see books, even at the elementary level, address computational issues. Some of them come even with a floppy disk containing specific programs and data sets. The process of learning to use these programs is sometimes painful, but in any case, it takes too much time. Also, it is too often uneasy because of an "unfriendly" system environment. Moreover, the topics which can be illustrated are not always those for which the use of a personal computer is most beneficial.

Another solution is to use one of the powerful statistical packages which are now available for personal computers. The catch is the high price which is justified by the professional use of these programs, but which cannot be absorbed at the level of a classroom or an individual. Cheap student versions are sometimes available, but, like their professional counterparts, they suffer the same drawbacks: they have been designed for professional use and not in a pedagogical perspective. They are too often ill prepared for class experiments.

The ultimate goal is clear: enhancement of the learning process with the use of interactive tools, solid mathematics, real life problems and up to date scientific concepts and technology. Mathematics and computer environments can be beneficially taken advantage of to learn how to model and solve the important statistical problems of real life.

The authors quickly realized that the implementation of these changes could not be done without the development of the appropriate tools: lecture notes and computer programs. The software package developed by the authors makes modern graphical and computational intensive procedures accessible to many and in particular to undergraduate students. It also includes a large number of examples.

The enthusiasm of the several colleagues and of the students first exposed to this new material lead the authors to consider the publication of the pedagogical materials they developed. But before we present these course materials, we include a discussion of a few (standard) thoughts about the use of computer programs in the learning process of statistics.

1.1. The Use of Statistical Packages

The use of statistical computing packages has been identified as the source of serious problems. This is particularly the case:

 a) when the users lack the knowledge to use a package properly,

 b) when instructors take any significant fraction of time from statistics courses to teach the use of the packages,

 c) when nonstatisticians imply that the packages can substitute for a knowledge of the underlying statistical methods.

Points a) and c) are well known side effects of the wide spread use of statistical packages in most of the branches of the scientific activity. Their effect can be disastrous and a serious effort is needed to make sure that the students we are preparing will not *fall into the pits.* But when it comes to point b) we can disagree and remark that the teaching of any applied statistical technique must include a demonstration of how it should be performed. Some statistical packages make such a task easy and instructive (see Velleman (1989), Härdle (1990), Tierney (1990), Nummi *et al.*(1989)). And there is nothing to fear as long as the first exposures to the computer packages take place in the classroom under the direction of trained statisticians. In fact the theoretical knowledge acquired from the lectures and the texts is not worth much in practical situation if the know-how is not there. Moreover *many students find it easier to understand the what and when of statistics when there is a complementary focus on the how.*

1.2. The Course Materials

In order to set up the pedagogical experiment described above, the authors developed the following course materials.

1.2.1. Two Textbooks

The lecture notes prepared by the authors are being polished and their material is reorganized for publication in the form of two textbooks.

Mathematical Statistics: a Computational Approach

This first book is intended as a textbook for an undergraduate (upper division to be specific) course in Mathematical Statistics. The authors had to face the difficult problem of simultaneously satisfying both those students aiming for further statistics courses and those for which this will be the last statistics course. The topics covered contain those of many other first year and upper division existing textbooks.

Univariate descriptive statistics featuring the now standard use of many of Tukey's exploratory data analysis techniques are presented first. This first chapter is succeeded by the usual chapters on distributions and random variables. Also classical are the topics of normal theory hypothesis testing and confidence intervals which in turn give way to analysis of variance and simple regression. Although these topics are familiar enough, many things distinguish this book from its competitors. Most of the examples and problems are described with actual (or simulated) computer manipulations using the program. Three chapters are devoted to the generation of pseudo-random numbers and random simulation. One chapter is devoted to the meaning and the applicability of limit theorems, one chapter is devoted to resampling methods and the bootstrap, and a final chapter is devoted to the notion of influence and to robust regression procedures. All this material is new at the level of the course. Some people were surprised that so much new material could be added to an already difficult undergraduate statistics curriculum. But the enlightenment of the software experiments (whether they are done by the instructor during the lectures or the students for their homework) made this whole operation possible and successful.

Each section ends with a list of illustrative problems and computer experiments. Both the problems and the computer experiments are an integral part of the learning process. Obviously, the computer experiments could be performed with the help of existing statistics software packages. But the technical difficulties inherent to learning how to use these packages and/or the inadequacy of the graphical interface and/or the price ... discourage the authors from relying on any of these existing programs.

Regression and Smoothing

This second textbook can be regarded as a sequel to the first one. It is intended for a first year graduate course. The first part of the material is fairly standard: multivariate normal distributions, introduction to the theory of linear models, review of the estimation theory and the tests for normal families, and an important chapter on multiple (least square) linear regression. But this text also contains a certain amount of nonstandard material. A long chapter on smoothing is new (at least at the level of the course). Locally linear smoothers are studied: lowess, super-smoothers, smoothing splines, kernel smoothing. An introduction to cross-validation is given in the process.

1.2.2. A Macintosh Program

In order to illustrate the theoretical concepts introduced in the lectures, the authors developed a specific interactive computer program.

The program has been used during the lectures as well as in the lab equipped with individual Macintosh stations. Obviously, the instructor can use a projection system to take advantage of the program during the lectures. This method was used systematically during the experimentations of this new teaching program and the results were extremely encouraging. The students like to see what is going on immediately without having to wait to go home or to the lab. Moreover, they seemed to be more likely to ask questions. And if the program has been especially designed for the experiment in question, the students are not side-tracked by irrelevant problems and their curiosity is rightfully focused on the concepts of interest.

The program was written for Macintosh computers, in two versions: one version for Macs with a coprocessor and one for Macs without a coprocessor (Plus, SE or Portable). The rationale is simple: the authors wanted to make sure that the training will be minimal. It has been proved in several instances that instructors and teaching assistants can learn to use the program in no time, in fact almost as fast as the students !!! Moreover, more than because of its friendly interface, Macintosh computers were chosen because of their graphical capabilities.

2. Typical Examples of Experiments

This section is devoted to the illustration of our system by the presentation of a few selected examples of experiments. We are tackling a problem which is common to all papers in this area: we are using static means to show how dynamic procedures work. The program uses three different windows: the table view window, the graphics window and the report text window. It is in the table window that data is first entered, where it is edited, and where the results of transforms are best seen. Each variable is shown as one of the columns of the table and each case of the data is shown as a row. Graphs are displayed automatically in the graphics window. Statistical output is displayed in the report text window. The text in the report window is fully editable. The students might use this to delete extraneous material or to add notes, when preparing their homework.

2.1. Graphical Estimation and Goodness of Fit

Superposing graphics

A great deal of time and space is devoted to the various methods of generation of random samples with a given distribution. It is enough to specify the sample size and the distribution to obtain such samples. But the program offers the possibility to generate samples of pseudo random numbers by following the steps of the most commonly used methods of generation: inversion of the cumulative distribution function, rejection methods, etc .... It is very easy to illustrate the first of these methods of generation. Indeed, the cummulative distribution functions of the usual probability distributions and their inverses are part of the set of functions which can be used for the "calculus" on the columns of the spreadsheet. The experiments go like this:

Experiment 2.1. *Generate a column (univariate dataset) of size N, say $N = 500$ from the uniform distribution. Create a new column by computing (entry by entry) F^{-1} for the distribution function of the sample we want to create. Plot a histogram, a smooth estimate of the density (a kernel estimate for which the user can choose the bandwidth if he/she so desires). The student can also use goodness of fit tests to check that the sample so-obtained is satisfactory. But the nicest feature of all is the possibility to superimpose, on the same graph as the histogram or the estimate of the density, the theoretical density of the distribution in question (see Fig. 2.1). This graphical option is very helpful and the students love it. Especially at the beginning of the year when they are not yet familiar with the concepts of test and p-value.*

Histograms

The graphic capabilities of the program make possible a very interesting analysis of the possibilities and the limitations of the notion of histogram. It has been recognized that, though

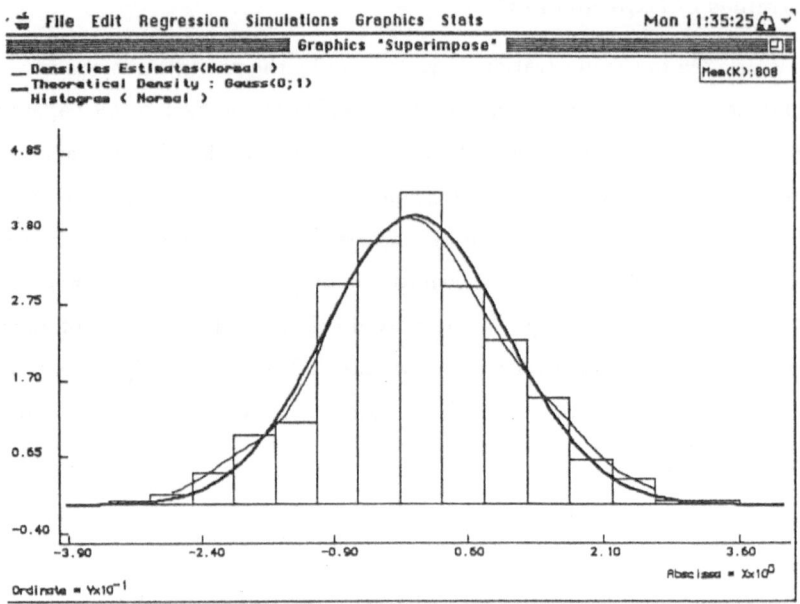

Figure 2.1: *The plot displays a kernel density estimate, a histogram estimate and superimposed the density of the standard normal distribution (thick line). The plot was obtained by specifying the appropriate options in the corresponding dialog window.*

extremely useful, histograms can be very misleading (see Härdle (1990)). The extreme sensitivity of histograms to the choice of the number of bins and their locations is rarely mentioned to the students and illustrated in a convincing way. The program is designed in such a way that the user can choose the binwidth and the leftmost point of the histogram. In this way it is easy to provide the students with examples for which the histogram looks unimodal for one choice of its parameters and trimodal for another choice. The Fig. 2.2 displays the plots of four histograms computed on the same data set. They all look different although their only difference is the shifted origin.

2.2. Limit Theorems

The law of large numbers can be illustrated in a very natural way. Here are the various steps to follow in order to perform such an experiment with the program. We give all the technical details to show that the manipulations which are left to the students are only here for pedagogical reasons.

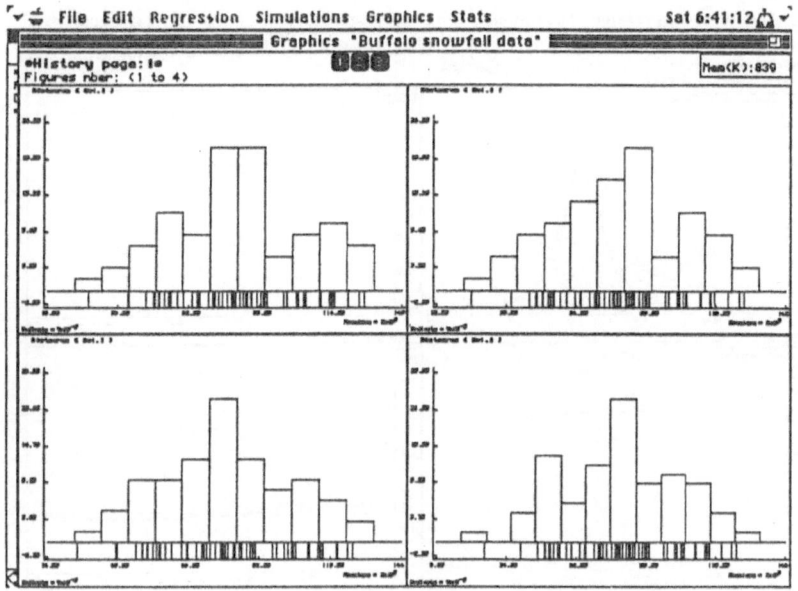

Figure 2.2: *Four histograms for the same dataset with the same binwidth $h = 10$, but with different origins $x = 2, 4, 6, 8$.*

Experiment 2.2. *Create a new dataset. To do so generate a column of size n, say $n = 500$ from the uniform distribution. Let X_i be the entries of this column. Use the algebraic manipulation available in the Editor window to create the columns corresponding to S_n and $\leq X = S_n/n$. After going back to the main menu, select the item **Data** from the **Graph** menu, and make a sequential plot of the sample so-obtained of $\leq X$. In order to see the convergence one can superimpose the straight line $y = 0.5$ on the graph (see Fig. 2.3).*

This experiment can be repeated with different distributions. It is possible to convince the student that the convergence has to take place, but when the Cauchy distribution is used, things are different. The sequence of normalized partial sums "tries" to converge toward 0 but it is repeatedly hit by a very large value which drives the sequence very far away from 0. This phenomenon is strikingly different from what was first observed. Without trying to prove anything, this experiment shows that not all the distributions are equal when it comes to the convergence of S_n/n. Once the curiosity of the students has been excited, it is possible for the instructor to reflect on the assumptions of the mathematical result and explain in this particular case that the existence of a first moment may be playing an important role. Now it is the right time to talk to the students about the existence of hypotheses for all these mathematical theorems, and that it is not a good idea to try to tamper with them (see Fig. 2.3).

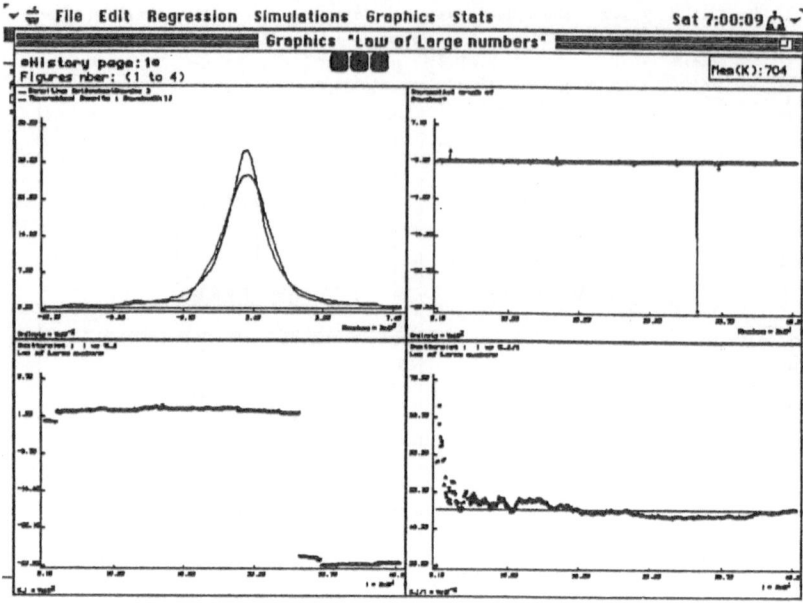

Figure 2.3: *An example illustrating the law of large numbers for a uniform sample and its failure for the Cauchy distribution. The top-left figure displays a kernel density estimate of the Cauchy sample which is sequentially displayed on the plot at the top-right figure. For this Cauchy sample the sequential plot of the normalized partial sums is displayed at the bottom-left graph. The last plot displays the sequential plot of the normalized partial sums for a sample from the uniform distribution.*

The central limit theorem is the object of a special item in the menu. This item is designed to illustrate the notion of convergence in distribution. Also, the Cauchy distribution can be used to show that the normalized partial sums are not always (approximately) normal and that it is sometimes beneficial to change the square root normalization and obtain a nontrivial limit under unexpected circumstances. The students seem to understand these subtleties without any difficulties. They do not have to take the word of the instructor for it: they see it!

2.3. Random Number Generation

Pseudo random numbers can be studied and generated in a such a way that all the steps of the creative process are under the control of the student.

We have already mentioned the possibility of using the inversion of the distribution function to generate random samples from a given distribution. A detailed discussion of the rejection method is given in the text. Many examples are discussed and the instructor can illustrate some, if not all of them, in class with the program. Indeed, the logical operations on the columns make it extremely simple to generate instantaneously, large samples from a given distribution if a rejection algorithm exists for it.

For multivariate (Gaussian) variables the simulations are done with the use of a set of matrix manipulation tools which make it possible to compute inverses and square roots of covariance matrices.

2.4. Bootstrap experiments

Resampling methods are used to estimate the sampling distribution of a statistic by taking random samples from the data. They are attractive as they rely on the substitution of computational power for theoretical analysis. We now describe how to perform the simplest example of the bootstrap methodology.

Experiment 2.3.   *Suppose that one needs a 90% confidence interval for the mean θ of a Gaussian distribution with unknown variance, having observed a sample of size 100. The first thing to do, is to compute the mean and the standard deviation of the sample, say $\hat{\theta} = 15.94$ and $\hat{\sigma} = 1.96$. Then use Monte Carlo simulations to produce bootstrap replications for the estimate of the mean. You can do this by setting the number of bootstrap replications B to a number, say 300, producing a column containing the 300 bootstrap replications of the mean. For computing the simple bootstrap pivotal interval, determine the 5% and 95% percentiles of the bootstrap distribution of the numbers contained in this column. By subtracting these percentiles to $2\hat{\theta}$ one obtains the simple bootstrap interval.*

2.5. Influence and Outliers in Simple Regression

The notion of outliers and influent points is very esoteric for undergraduate students. The notion of breakdown point is even more difficult to grasp. In fact these topics are, most of the time, ignored at this level. The use of the program designed by the authors offers an unprecedented opportunity to familiarize the students with these concepts. Diagnostic index plots of studentized residuals, Welsh-Kuh's distance, etc arranged in a rectangular array can be displayed on the screen (see Fig. 2.4).

During a regression analysis, the user deletes, adds, or moves data points on the screen. Simultaneously he/she sees the effect of his/hers action on the regression line which is updated as the data are changed. Fig. 2.5 illustrates what happens on the screen when doing these operations. Comparison with the old one is possible since this old regression line still appears in a lighter tone. The program takes advantage of the superiority of dynamic plots over listing outputs in studying regression diagnostics.

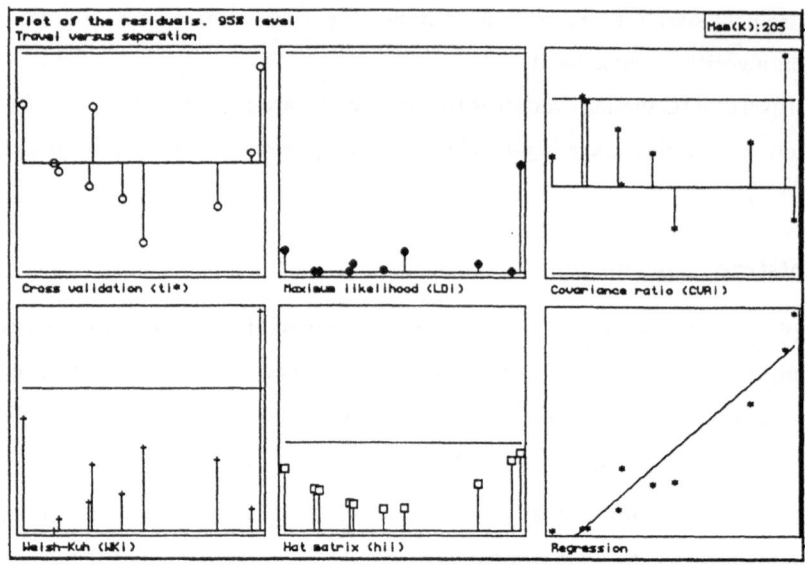

Figure 2.4: *Diagnostic index plots of studentized residuals, Welsh-Kuh's distance, maximum likelihood distance, diagonal elements of the hat matrix, etc arranged in a rectangular array can be displayed on the screen.*

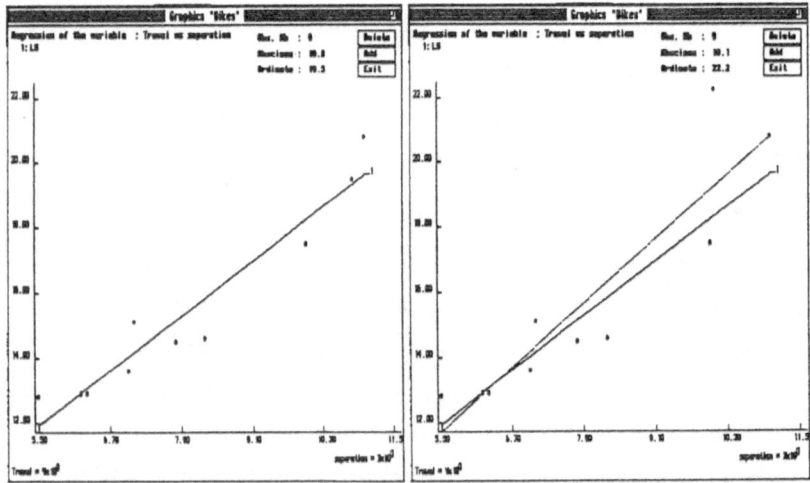

Figure 2.5: *A regression line fit. The data setis the bicycle data set of Devore and Peck (1986) (see also Tierney (1990)). A point has been moved from its original position. The left figure displays the regression fit of the original data.*

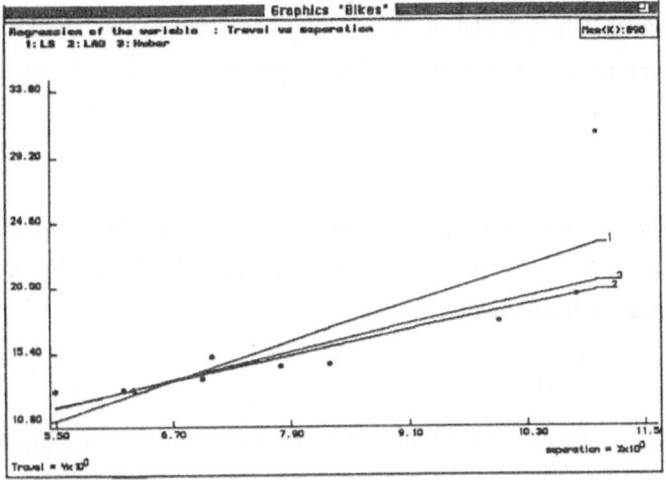

Figure 2.6: *A comparison plot for comparing several estimation methods on the same dataset.*

2.6. Robust Regression

Motivated by experimental analysis (and a working knowledge) of the notion of influence and breakdown points, the students find the study of robust alternatives to the least square regression very natural. Least absolute deviations and least median of squares regressions, Huber method, ... are possible choices for a simple regression. Several methods can be used on the same data set and the graphs of the various regression lines can be plotted simultaneously (see Fig. 2.6).

Together with the graphical analysis of influent points which we discussed earlier, these plots give the right feeling about robust regression.

3. Conclusion

We beleive that the changes in the undergraduate teaching of statistics which we described above are unavoidable. They will be imposed, sooner or later, by the needs of a modern economy.

Something has to be said in favor of the implementation we proposed. The students participating in the experiments had a heavier working load. They had to learn more than the students of the regular curriculum. They seemed to be doing fine. In fact, the pedagogical experiment was continued and expanded everywhere it was tried. We take this as a sign of success.

At the time of the preparation of this note, the manuscripts of the two textbooks and the manual and the program itself are still being worked on. They will be published by Prentice

Hall. They should be available in the Fall of 1992 or early in 1993. The current version of the program is available for review upon request. The requests should be addressed to René Carmona.

Acknowledgments

We wish to thank Alain Filhol from the Institut Laue Langevin of Grenoble for his help and advices in implementing the code of the software while it was being developed. We also wish to thank an anonymous referee for helpful comments.

References

1. *Devore J. and Peck R. (1986). Statistics, the exploration and analysis of data, West Publishing Company, Mn.*

2. *Härdle W. (1991). Smoothing Techniques with implementation in S, Springer-Verlag series in Statistics, New York.*

3. *Nummi T., Nurhonen M. and Puntanen S. (1989). Dynamic illustration of Regression diagnostics, I.S.I.–47th Session, Paris, pp. 33–49.*

4. *Tierney L. (1990) . LISP-STAT An object-Oriented Environment for Statistical Computing and Dynamic Graphics, John Wiley & Sons, New York.*

5. *Velleman P. F. (1989). Learning data analysis with Datadesk, Freeman & Company, New York.*

Bayesian Electromagnetic Imaging

M. Roussignol[1], V. Jouanne[2], M. Menvielle[3], P. Tarits[2]

[1] Laboratoire de Statistique et Probabilités, U.F.R. de Mathématiques, Université des Sciences et Technologies de Lille, 59655 Villeneuve d'Ascq Cedex, France.
[2] Laboratoire de Géomagnétisme, Institut de Physique du Globe de Paris, Tour 14, 4 place Jussieu, 75252 Paris Cedex 05, France.
[3] Laboratoire de Géomagnétisme, Institut de Physique du Globe de Paris, Tour 14, 4 place Jussieu, 75252 Paris Cedex 05, France.
now at :
Laboratoire de Géophysique, Bâtiment 504, Université Paris Sud, 91405 Orsay Cedex, France.

Key words : Electromagnetism, Inverse Problem, Bayesian Statistic, Stochastic Algorithm.

Abstract

This work presents a method to find rock conductivities in a zone from electromagnetic measurements on the surface of the earth. It uses a stochastic algorithm to find a Bayesian estimator of the conductivities. The algorithm is tested on a synthetic model made up of an heteregeneous thin sheet inbedded in a stratified substratum.

1. Introduction

The ultimate purpose of electromagnetic induction studies is to estimate the distribution of rock conductivities from electromagnetic measurements, i.e. to find a solution of the inverse electromagnetic problem. The measurements are made at stations more or less evenly distributed on the surface of the earth, in a limited frequency range and with experimental noise.

Reaching this target first requires to solve the forward electomagnetic induction problem, i. e. to calculate electric and magnetic fields on the surface when the distribution of rock conductivities beneath surface is known. This is achieved mostly when geometrical particularities of the distribution of conductivities induce simplifications in the computation and in turn significant savings of computer time.

It is for instance the case when the conductivity does not depend on one horizontal direction or when the lateral heterogeneities of the conductivity are located in a thin layer embedded in a stratified substratum (thin sheet approximation, see Mareschal and Vasseur 1984 for a review of the already proposed solutions). Vasseur and Weidelt (1977) have proposed a solution of the forward problem with a heterogeneous thin sheet on the surface of a stratified medium and showed that their formalism mat also be used to solve the inverse problem. Barthes and Vasseur (1980) have given an example of inverse problem solution using least-square optimisation technic.

This paper aims at presenting a bayesian general method to solve the inverse problem, when a solution of the forward problem is available. We search for solutions to the distribution of conductivities in a restricted class of conductivity, either on account of knowledge on the rocks or on account of simplification choices. The Bayesian method first expresses these knowledge and choices by a probability law on the distribution of conductivity, named a-priori law. Then it calculates the a-posteriori law, a probability law on the distribution of conductivities which takes into account the observations and the probability density of the experimental noise. From the a-posteriori law, it is possible to calculate a mean value which gives an estimator of the distribution of conductivities and a variance which gives an indicator of the precision of the estimation. Bayesian technics in geomagnetism have already been studied and used either with gaussian a-priori laws and gaussian experimental noise (*Backus 1988, Mackie and al. 1988*) or with uniform a-priori laws and gaussian experimental noise (*Tarits and al. 1991*).

When we leave out the gaussian hypothesis, the analytical formula for the a-posteriori law implies a computation impossible to conduct directly. So, with general a-priori and noise laws, we use a stochastic algorithm which is composed of the simulation of a Markov chain being stabilized on the a-posteriori law. This algorithm is close to "heat bath" and "simulated annealing" algorithms used in imaging (*Kirkpatrick 1984, Geman and Geman 1984, Gibert and Virieux 1991*).

Basic equations are given in section II. We calculate in section III the a-posteriori law with general a-priori and noise laws. We discuss different kinds of a-priori law and give a formula with a gaussian noise law. In section IV we describe a Markov chain which is stabilized on the a-posteriori law. Then we give in section V a modified Markov chain whose simulation saves computer time compared to the basic Markov chain. Finally we give in section VI the results of an experiment with a synthetic model of heteregeneous thin sheet.

2. Basic equations

Let us assume that we are able to calculate temporal Fourier transforms $E_n(M,f)$ and $H_n(M,f)$ of the electric and magnetic field at any point M of $\mathbf{R}^3$ and any frequency f with a given structure, named normal structure. $E_n(M,f)$ and $H_n(M,f)$ are 3D complex vectors. The normal structure is defined by the conductivities $\sigma_n(M)$ at any point M of $\mathbf{R}^3$. The quantity $\sigma_n(M)$ is a positive number. For instance normal structure can be a sequence of uniform horizontal layers. The real structure differs from the normal structure in a domain D of $\mathbf{R}^3$. Let us call $\sigma(M)$ the real number which measure the conductivity at point M for the real structure and σ the function from $\mathbf{R}^3$ into $\mathbf{R}_+$ which represents the real structure. We assume that $\sigma(M)$ and $\sigma_n(M)$ are equal when M is outside the domain D of interest. Abnormal conductivity $\sigma_a(M)$ at point M denotes the difference $\sigma(M) - \sigma_n(M)$ between real conductivity and normal conductivity at point M. $E(M,f,\sigma)$ and $H(M,f,\sigma)$ denote temporal Fourier transforms of the electric and magnetic field at point M and frequency f with the real structure σ. $E(M,f,\sigma)$ and $H(M,f,\sigma)$ are 3D complex vectors. Using the Green kernel method, we can rewrite basic equations of electomagnetism as (*Weidelt 1975*) :

$$E(M,f,\sigma) = E_n(M,f) - i \left(f\mu \int_D \sigma_a(N) \, G(M,N,f) \, E(N,f,\sigma) \, dN \right)$$

$$H(M,f,\sigma) = H_n(M,f) + \int_D \sigma_a(N) \, rot(G(M,N,f)) \, E(N,f,\sigma) \, dN$$

at any point M of $\mathbf{R}^3$ and any frequency f. The quantity μ is the magnetic permeability of medium. We will consider it as equal to that of the vacuum everywhere. $G(M,N,f)$ is a 3×3 complex matrix, named Green kernel. $G(M,N,f) \, \delta(N)$ represents the Fourier transform at frequency f of the electric field at M created by a unit dipole $\delta(N)$ located at N (*Morse and Feshback 1953*). $rot(G(M,N,f))$ is a 3×3 complex matrix obtained by taking the rotational in M of the field of the column vectors of the matrix $G(M,N,f)$. The integrals in the formulas are integrals in the variable N of $\mathbf{R}^3$ over the domain D.

To solve these equations, we discretise the domain D in K small cells P_k $(k = 1,...., K)$ in which conductivities, Green kernels, electric and magnetic fields are nearly constant. For instance the cells P_k are little cubes the union of which is the domain D. Then basic equations become, using same notations for the sake of simplicity and denoting $\tau = (\tau(P_k) ; k = 1, ..., K)$ the abnormal integrated conductivities of the cells P_k (i. e. the conductances $\int_{P_k} \sigma_a(N) \, dN$) :

(1) $$E(M,f,\tau) = E_n(M,f) - i \left(f\mu \sum_{k=1}^{K} \tau_k \, G(M,P_k,f) \, E(P_k,f,\tau) \right)$$

(2) $$H(M,f,\tau) = H_n(M,f) + \sum_{k=1}^{K} \tau_k \, rot(G(M,P_k,f)) \, E(P_k,f,\tau)$$

at any point M and any frequency f.

Let us consider the linear system made up with equation (1) written at each point $M = P_j$ $(j = 1,....,K)$ in D. Given the abnormal conductance $\tau = (\tau(P_k), k = 1,..., K)$, the quantities $E(P_k,f,\tau)$ are solutions of this system. For each frequency, the system has 6 K equations and 6 K real unknowns in the general case. Then the electric field $E(M,f,\tau)$ and the magnetic field $H(M,f,\tau)$ at any point M and any frequency f are calculated using (1) and (2).

Let us consider now the following problem. The conductances $\tau = (\tau(P_k)$; $k = 1,...,K)$ are unknown and we want to estimate them with measurements at different points of the surface and at different frequencies. We name this problem the inverse problem. The data $y_{i,j}$ are measurements of a known function $h(E(M_i,f_j,\tau),H(M_i,f_j,\tau))$ at given points M_i $(i = 1,....,I)$ on the surface and for given frequencies f_j $(j = 1,......,J)$. For each point M_i and each frequency f_j, the data $y_{i,j}$ is equal to $h(E(M_i,f_j,\tau),H(M_i,f_j,\tau))$ plus an experimental noise $w_{i,j}$:

$$y_{i,j} = h(E(M_i,f_j,\tau),H(M_i,f_j,\tau)) + w_{i,j}$$

We suppose that the data $\mathbf{y} = (y_{i,j}$; $i = 1,...,I$; $j = 1,...,J)$ and the noise $\mathbf{w} = (w_{i,j}$; $i = 1,...,I$; $j = 1,...,J)$ are realisations of random variables $\mathbf{Y} = (Y_{i,j}$; $i = 1,...,I$; $j = 1,......,J)$ and $\mathbf{W} = (W_{i,j}$; $i = 1,...,I$; $j = 1,...,J)$. We suppose that the random variables $(W_{i,j}$; $i = 1,...,I$; $j = 1,...,J)$ are independent and that for each point i and each frequency j they have the probability density $p_{i,j}$. The random variables are supposed to have 0 as mean value. Usually it is assumed that these densities are gaussian, but it is not necessarily always true.

Solving the inverse problem consists in estimating the conductances $\tau = (\tau(P_k)$, $k = 1,...., K)$ given the data $\mathbf{y} = (y_{i,j}$; $i = 1,....,I$; $j = 1,....,J)$.

3. Bayesian estimation of the unknown structure

We are searching for solutions to the distribution of condutances in a restricted class of conductances, either on account of geophysical knowledge or on account of simplification choices. The geological knowledge comes from previous investigations in the studied aerea (geology, other geophysical methods...) and from laboratory measurements of rock conductivity. The Bayesian method expresses these knowledge and choices by a probability law q on the set of possible values of conductances, named a-priori law. The support of q is the first choice. Maximum and minimum values of possible conductances on each cell are deduced from results from other earth sciences. Numerical computations make it necessary to define the support of q as a finite set. In what follows, we take L different values $c_1,....,c_L$ of possible conductances on each cell. The support of q is then the set A of the L^K possible elements $\mathbf{a} = \{a_k$; $k = 1,..., K\}$ where the a_k belong to $\{c_1,...., c_L\}$. We name each element of A an image of conductance. As usual, the greater the L, the more precise the model, but the longer the computation. We take the same set of L possible conductances for each cell. But in any cell, it is possible to limit the search of conductances in a smaller range of value by making the a-priori probability of the other values equal to zero. If no other information is available, we take as a-priori probability the uniform probability on the range of admitted images of conductance. In some geophysical situations, we can notice that anomalous conductances smoothly vary with space, i. e. without too important oscillations. The image has therefore to fit with some smoothness requirements. It can be obtained by taking the a-priori law as a "Gibbs probability" (*Geman and Geman 1984*) which gives a great probability to smooth images and little probability to other images. For instance we can take the following probability :

$$p(\mathbf{a}) = Z \, exp\left(-\lambda \sum_{k_i \text{ and } k_j \text{ neighbours}} (a_{k_i} - a_{k_j})^2 \right)$$

where Z is a normalizing constant, λ a positive parameter and $\mathbf{a}$ an image of conductance. The λ parameter drives the expected smoothness of the distribution of conductances in D. The greater λ is, the smoother expected distribution of conductances; if λ equals 0, there is no a-priori infomation about smoothness.

From the Bayesian point of view (*Tarantola 1987, Press 1989*), both observations and conductances are random variables. The law of the observations is the conditional probability given the values of conductances whose density is given by :

(3) $$f(\mathbf{y},\mathbf{a}) = \prod_{i=1}^{I} \prod_{j=1}^{J} p_{i,j}(y_{ij} - h(E(M_i,f_j,\mathbf{a}),H(M_i,f_j,\mathbf{a})))$$

This quantity is a function of $\mathbf{a} = (a_k ; k = 1,.....,K)$ through E and H (see equations (1) and (2)). It is a density of probability for the variable $\mathbf{y} = (y_{i,j} ; i = 1,..,I ; j = 1,...,J)$.

If the probability densities p_{ij} are gaussian, formula (3) can be written :

$$f(\mathbf{y},\mathbf{a}) = Z_1 \, exp\left(-\sum_{i,j} \frac{(y_{ij} - h(E(M_i,f_j,\mathbf{a}),H(M_i,f_j,\mathbf{a})))^2}{2 (\sigma_{ij})^2} \right)$$

where Z_1 is a normalizing constant.

We are interested in the a-posteriori law for the conductances, i. e. the conditional probability of the conductances given the observations. Standard computation of conditional probabilities gives the a-posteriori law for any image $\mathbf{a}$ of conductances and any data $\mathbf{y}$:

(4) $$P(\tau = \mathbf{a} / Y = y) = \frac{f(\mathbf{y},\mathbf{a}) \, q(\mathbf{a})}{\sum_{b \in A} f(\mathbf{y},\mathbf{b}) \, q(\mathbf{b})}$$

We take the mean value of the a-posteriori probability as an estimator of the distribution of conductances . If $A(k,c_j)$ is the set of images that have the conductance c_j in the cell k, the k-marginal a-posteriori probability po_k is :

$$po_k(c_j) = P(\tau \in A(k,c_j) / Y = y) = \frac{\sum_{\mathbf{a} \in A(k,c_j)} f(\mathbf{y},\mathbf{a}) \, q(\mathbf{a})}{\sum_{\mathbf{a} \in A} f(\mathbf{y},\mathbf{a}) \, q(\mathbf{a})} \quad .$$

The estimator of the conductance of the cell k is the mean value of the k-marginal a-posteriori
probability : $\sum_{j=1}^{L} c_j\ po_k(c_j)$.

The variance of this marginal probability gives an indication of the validity of the estimation. Another indication is to draw the diagrams of all the marginal probabilities to see if they are univariate and if the mean value is an acceptable estimator. The a-posteriori probability incorporates two factors of imprecision : the noise on the measurements and the non-unicity of the inverse problem.

The function h(.,.) may be chosen in different ways. In the case of magnetotelluric soundings, it is the impedance tensor. This tensor may be expressed on the observation bases, or on a couple of real or complex basis (one magnetic basis and one electric basis) on which the tensor is antidiagonal (*Eggers 1982, Counil and al. 1986, La Torraca and al. 1986*). In the case of geomagnetic sounding, it is in general the tipper which relates the vertical component of the magnetic field to its horizontal component. In some cases, it is interesting to use both the impedance tensor and the tipper. The function h is calculated with the quantities $E(M_i, j_j, \tau)$ and $H(M_i, f_j, \tau)$, each computation generally requiring to solve J linear systems of 6 K equations and 6 K unknown (see section 2).

In the quite simple formula (4), there is a hidden difficulty : the computation of the sum in the denominator. This computation requires to calculate f(y,b) q(b) for all the possible images **b** of conductances, i.e. L^K times. In real conditions, this is impossible to achieve directly. So we will use an algorithm which permits to calculate the a-posteriori law.

4. Markov chain algorithm

We are going to define an algorithm which simulates a Markov chain stabilized on the a-posteriori law. Let us denote $\tau^{(n)} = (\tau^{(n)}(P_k), k = 1,..., K)$ the image of conductances obtained at step n of the algorithm. We start from an initial image of conductances $\tau^{(0)}$. At step n, we update the image $\tau^{(n)}$ by changing the image into only one cell. We choose the cell by a random selection with the uniform probability over the K cells. If this cell is the k-th, the new value of the conductance at this cell is chosen by a random selection with the following probability law :

$$P(\tau^{(n+1)}(P_k) = c_j) = \frac{f(y,\ \mathbf{a}(\tau^{(n)},k,c_j))\ q(\mathbf{a}(\tau^{(n)},k,c_j))}{\sum_{j=1}^{L} f(y,\ \mathbf{a}(\tau^{(n)},k,c_j))\ q(\mathbf{a}(\tau^{(n)},k,c_j))}$$

where $a(\tau,k,c_j)$ denotes the image equal to τ on all the cells other than k-th and equal to c_j on the cell k. The computation of $P(\tau^{(n+1)}(P_k) = c_j)$ requires to calculate f(y, $\mathbf{a}(\tau^{(n)},k,c_j)$) q($\mathbf{a}(\tau^{(n)},k,c_j)$) L times.

The sequence of images $(\tau^{(n)})_{n \geqslant 0}$ is a random process which is an homogeneous Markov chain : the law of $\tau^{(n+1)}$ depends only on $\tau^{(n)}$ and on a random selection independent from the past. It is of the same kind as the Gibbs sampler (*Geman and Geman 1984*). This chain is irreductible and aperiodic over the finite space of all possible images. The a-posteriori probability is the invariant probability of this chain (see appendix for a proof). So according to the ergodic theorem for Markov chains, we have :

$$\lim_{n \to \infty} \frac{1}{n} \sum_{i=1}^{n} 1_{\{\tau^{(i)}(P_k) = c_j\}} = po_k(c_j) \quad a.s.$$

where $1_{\{\tau^{(i)}(P_k) = c_j\}}$ denotes the indicator function of the event $\{\tau^{(i)}(P_k) = c_j\}$.

This gives a method to estimate the marginal probabilities of the a-posteriori law and their expected values and variance.

In this algorithm, we must solve at each step $L \times J$ linear systems of dimension 6K to calculate the electric field in D for each possible next image of conductance and for each frequency of measurement. The systems can be solved by an iterative method of resolution like Gauss-Seidel method. We take as initial point for each resolution the solution found at the end of the previous iteration. The number of steps necessary to make relevant statistics on the stationary state is about $K \times 200$. So there are many computations to do. However it is more conceivable than to compute the $L^K \times J$ linear systems of dimension 6K required for a direct computation of the a-posteriori law.

5. Modified Markov chain

We are about to make a modification of the above method to save computer time. At each step, we update the image of conductances like above but, instead of solving the linear systems, we also update the electric field on D by one iteration of some method like Gauss-Seidel.

We start from an initial image of conductances $\underline{\tau}^{(0)}(P_k)$ and an initial electric field $\underline{E}^{(0)}(P_k,f_j)$ on each cell P_k and each frequency f_j. At step n we denote $\underline{\tau}^{(n)}(P_k)$ the conductance on cell P_k, $\underline{E}^{(n)}(P_k,f_j)$ the electric field on cell P_k and for frequency f_j. We regularly scan the image and we denote $P_{k^{(n)}}$ the cell scanned at time n. Our algorithm is the following.

At step n we first update $\underline{E}^{(n)}$ to $\underline{E}^{(n+1)}$ on the cell $P_{k^{(n)}}$ for all frequencies f_j by one iteration of the Gauss-Seidel method :

$$\underline{E}^{(n+1)}(P_{k^{(n)}},f_j) = A^{-1} V$$

with : $A = I - \underline{\tau}^{(n)}(P_{k^{(n)}}) G(P_{k^{(n)}},P_{k^{(n)}},f_j)$

$$V = E_n(P_{k^{(n)}},f_j) - i (f_j \mu \sum_{l \neq k^{(n)}} \underline{\tau}^{(n)}(l) G(P_{k^{(n)}},P_l,f_j) \underline{E}^{(n)}(P_l,f_j))$$

Then we update the image $\underline{\tau}^{(n)}$ to $\underline{\tau}^{(n+1)}$ by changing the image only on the cell $P_{k(n)}$. $\underline{\tau}^{(n+1)}$ has the same values as $\underline{\tau}^{(n)}$ on all the cells except the cell $P_{k(n)}$. The new value of the conductance on this cell is chosen by a random selection according to the following probability law :

$$P(\underline{\tau}^{(n+1)}(P_{k(n)}) = c_j) = \frac{f^{(n+1)}(\mathbf{y},\, \mathbf{a}(\underline{\tau}^{(n)},k^{(n)},c_j))\, q(\mathbf{a}(\underline{\tau}^{(n)},k^{(n)},c_j))}{\displaystyle\sum_{l=1}^{L} f^{(n+1)}(\mathbf{y},\, \mathbf{a}(\underline{\tau}^{(n)},k^{(n)},c_j))\, q(\mathbf{a}(\underline{\tau}^{(n)},k^{(n)},c_j))}$$

where $\mathbf{a}(\tau,k^{(n)},c_j)$ denotes the image equal to τ on all the cells other than $P_{k(n)}$ and equal to c_j on the cell $P_{k(n)}$ and $f^{(n+1)}(\mathbf{y}, \mathbf{a})$ denotes :

$$f^{(n+1)}(\mathbf{y},\, \mathbf{a}) = \prod_{i=1}^{I}\prod_{j=1}^{J} p_{i,j}(\, y_{ij} - h(E^{(n+1)}(M_i,f_j,\mathbf{a}),H^{(n+1)}(M_i,f_j,\mathbf{a}))\,)$$

with : $E^{(n+1)}(M_i,f_j,\mathbf{a}) = E_n(M_i,f_j) - i\, f_j\, \mu_0 \sum_{k=1}^{K} a(k)\, G(M_i,P_k,f_j)\, E^{(n+1)}(P_k,f_j)$

$$H^{(n+1)}(M_i,f_j,\mathbf{a}) = H_n(M_i,f_j) + \sum_{k=1}^{K} a(k)\, rot(G(M_i,P_k,f_j)\, E^{(n+1)}(P_k,f_j)\,.$$

The sequence $(\underline{\tau}^{(n)}(P_k),\underline{E}^{(n)}(P_k,f_j)$; $k = 1,...,K$; $j = 1,...,J)$ is a random process on the space which is the product of the set of all possible images and the set $\mathbb{C}^{3KJ}$. It is a Markov chain because $(\underline{\tau}^{(n+1)}, \underline{E}^{(n+1)})$ depends on $(\underline{\tau}^{(n)}, \underline{E}^{(n)})$ and a random selection independent from the past. This chain is not standart for its state space is a product of a finite space and a continuous space. Experiments show that it behaves like an ergodic chain and that it permits to estimate the a-posteriori law as the previous algorithm does provide a relevant choice of the initial set $(\underline{\tau}^{(0)}, \underline{E}^{(0)})$. The mathematical proof of the convergence of the algorithm is not yet done. This new algorithm is much faster than the previous.

6. Experiment with synthetic models

We tested the algorithm described in section V with two synthetic models. The anomalous domain is a heteregeneous thin sheet embedded in a stratified substratum. For model 1a the thin sheet is on the surface whereas for model 1b the thin sheet is in depth at the bottom of the first layer (see figure 1a and 1b).

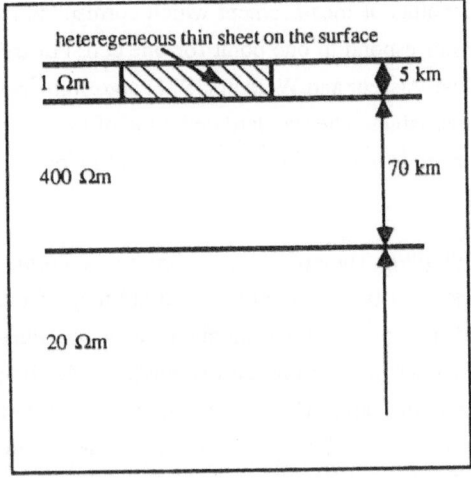

Figure 1a

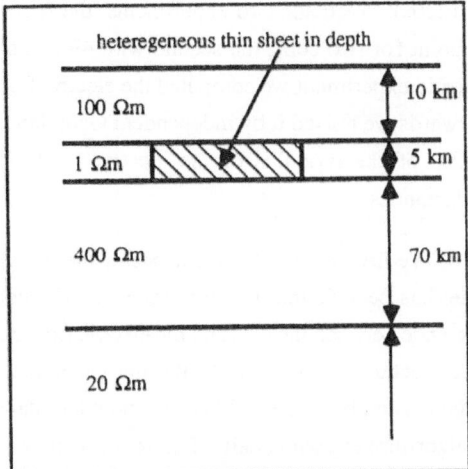

Figure 1b

The heteregeneous part of the sheet is in a square surrounded by an homogeneous domain which extends towards infinity in the two directions. It consists of 49 square shaped cells of 30km×30km. The heterogeneity resides in a resistant zone immersed in a conductive substratum. The distribution of conductances of the models we use is shown in figure 2.

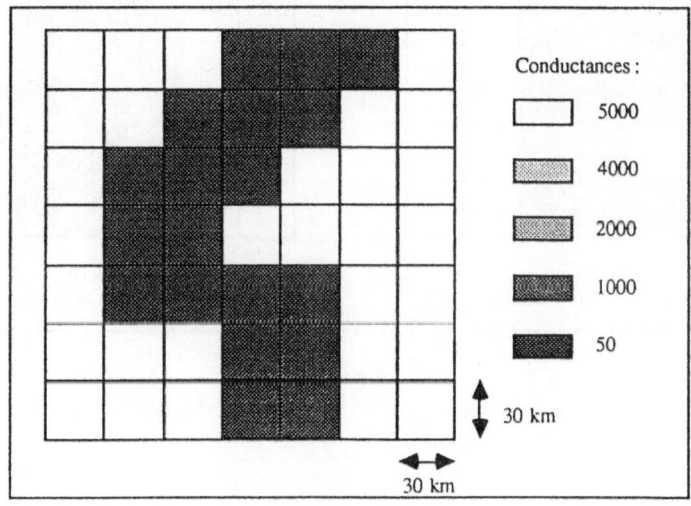

Figure 2

Observation points are evenly distributed on the surface above the abnormal zone. The data are the values of the electric field at the points of measurement for five frequencies (400s, 600s, 800s, 1000s

94

and 1200s). We made two experiments, the first with 25 points of measurement which correspond to one point for two cells, the second with 49 points which correspond to one point for one cell. For the synthetic experiment we computed the electric field with the Vasseur and Weidelt (1977) algorithm and afterwards we noised it by independent Gaussian random variables. The standard deviation of the noise is 10% of the modulus of the electric field. The a-priori law is the uniform probability over 5 conductances.

We investigated the influence of the initial set ($\underline{\tau}^{(0)}$, $\underline{E}^{(0)}$). The set corresponding to the normal model has been found the more relevant. The convergence can be measured by the amplitude of the difference between the data and the electric field computed at the points of measurements with the values of conductances found. With the normal model as initial set, this amplitude decreases in the first scannings and becomes stable after about ten scannings in all the cases. This proves the convergence of the algorithm experimentally. Figure 3 and figure 4 show the obtained images for models a and b with 25 points and 49 points of measurement.

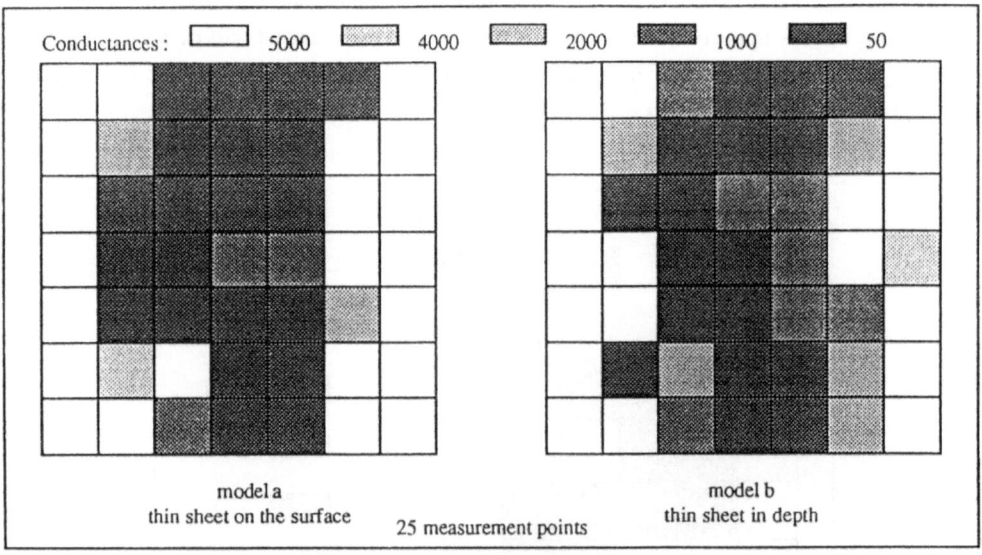

Figure 3

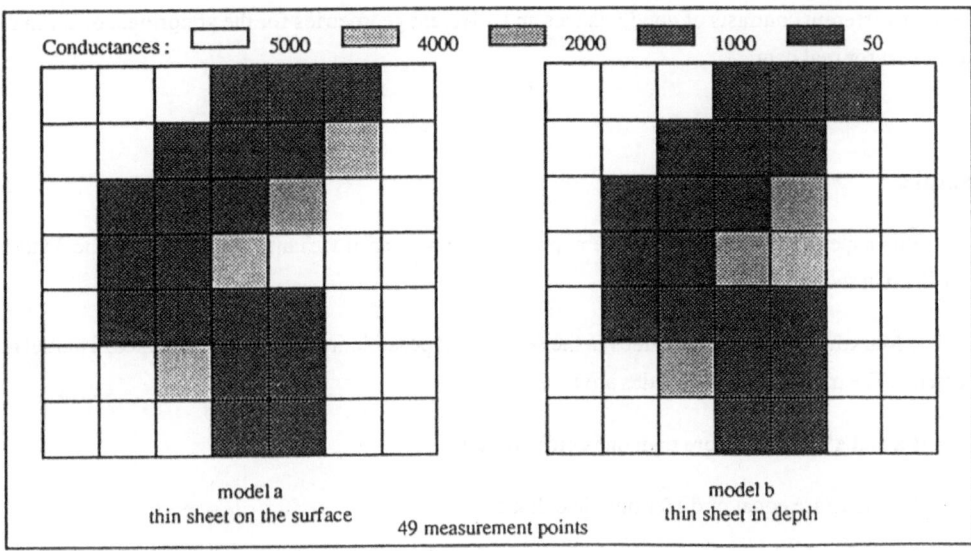

Figure 4

With 25 observation points (figure 3), the geometry of the abnormal zone appears, but with imprecisions on each side of the resistant structure. With 49 observation points (Figure 4), the conductance distribution is fairly well reproduced . The conductance estimates are exact except in narrow bands on either side of the resistive structure. These examples show that the stochastic algorithm almost exactly reproduces the conductance distribution when the size of the square cells is similar to the spacing of the measurement points.

7. Conclusion

The method presented here is general. It can work with any type of function h(.,.) for the data, any type of noise with known density of probability. It can incorporate some previous knowledge on the unknown structure. The method is available as soon as we are able to manage the direct problem. The main problem is that it leads to rather heavy computations, which are yet tractable with the modified algorithm.

We have tested it on a synthetic model for a heteregeneous thin sheet on the surface and in depth with the electric field as data and with two different densities of measurement points. In the case where the spacing of measurements points is similar to the size of the cells, the geometry of the abnormal zone was found again.

These results show the method is available, but there remains several works to be done. We must prove mathematically the convergence of the algorithm. We must test the algorithm with different kinds of data, with different contrasts of conductances and different geometries for the abnormal zone. Finally we must use it for real data.

Appendix

In this appendix, we prove that the a-posteriori law is an invariant probability for the Markov chain defined in section IV.

The Markov chain takes its values in the set A of all possible images of conductance. This set has L^K elements. The transition probabilities are :

- if $\mathbf{a}$ and $\mathbf{a'}$ differ in more than one cell : $P(\mathbf{a}, \mathbf{a'}) = 0$

- if $\mathbf{a}$ and $\mathbf{a'}$ are equal or differ only in cell k :

$$P(\mathbf{a}, \mathbf{a'}) = \frac{1}{K} \frac{f(y, \mathbf{a'})\, q(\mathbf{a'})}{\sum_{l=1}^{L} f(y, a(k,c_l))\, q(a(k,c_l))}$$

where $a(k,c_l)$ denotes the image equal to $\mathbf{a}$ or $\mathbf{a'}$ in all the cells different than the k-th and whose value in cell k is c_l.

A probability m on the set of images is invariant for the Markov chain if for all images $\mathbf{a}$ we have : $\sum_{\mathbf{a'}} m(\mathbf{a'})\, P(\mathbf{a'},\mathbf{a}) = m(\mathbf{a})$. The probability m is said reversible for the Markov chain if for all images $\mathbf{a}$ and $\mathbf{a'}$ we have : $m(\mathbf{a})\, P(\mathbf{a}, \mathbf{a'}) = m(\mathbf{a'})\, P(\mathbf{a'}, \mathbf{a})$. It is easy to see that any reversible probability is invariant.

The a-posteriori law is equal to :

$$m(\mathbf{a}) = \frac{f(y,\mathbf{a})\, q(\mathbf{a})}{\sum_{\mathbf{b} \in A} f(y,\mathbf{b})\, q(\mathbf{b})}$$

Let us show that the a-posteriori law is reversible.

If $\mathbf{a}$ and $\mathbf{a'}$ differs in more than one cell the relation of reversibility is obviously fulfilled. If $\mathbf{a}$ and $\mathbf{a'}$ are equals or differs only in the cell k, we have :

$$m(\mathbf{a})\, P(\mathbf{a}, \mathbf{a'}) = \frac{f(y,\mathbf{a})\, q(\mathbf{a})}{\sum_{\mathbf{b} \in A} f(y,\mathbf{b})\, q(\mathbf{b})} \frac{1}{K} \frac{f(y, \mathbf{a'})\, q(\mathbf{a'})}{\sum_{l=1}^{L} f(y, a(k,c_l))\, q(a(k,c_l))}$$

and this quantity is symmetrical in $\mathbf{a}$ and $\mathbf{a'}$, so is equal to $m(\mathbf{a'})\, P(\mathbf{a'}, \mathbf{a})$.

We suppose that for all images **a**, the quantity f(y, **a**) q(**a**) is strictly positive. So for any **a** and **a'** there is a strictly positive probability that the Markov chain will go from **a** to **a'** in K transitions. So the chain is irreductible. There is a strictly positive probability that in one transition the chain will stand in the same position. So the chain is aperiodic. Standard results on Markov chains show that the chain is recurrent, that there exists one and only one invariant probability and that the pointwise ergodic theorem is true.

References

Backus, G. E. (1988). Bayesian inference in geomagnetism. Geophysical Journal, 92, 125-142.

Barthes, V. and Vasseur, G. (1980). An inverse problem for electromagnetic prospection.

Counil, J.L., Le Mouel, J.L. and Menvielle, M. (1986). Associate and conjugate direction concepts in magnetotellurics. Annales Geophysicae, 4, B, 115-130.

Eggers, D.E. (1982). An eigenstate formulation of the magnetotelluric impedance tensor. Geophysics, 47, 1204-1214.

Geman, S. and Geman, D. (1984). Stochastic relaxation, Gibbs distribution, and the Bayesian restoration of images. IEEE Trans. Pattern Analysis and Machine Intelligence, 6, 721-741.

Gibert, D. and Virieux, J. (1991). Electromagnetic imaging and simulated annealing. J. Geophys. Res., 96, 8057-8067.

Kirkpatrick, S. (1984). Optimisation by simulated annealing : quantitative studies. J. Stat. Phys., 34, 975-986.

La Torraca, G.A., Madden, T.R. and Korringa, J. (1986). An analysis of the magnetotelluric impedance tensor for three-dimensional conductivity structures. Geophysics, 51, 1819-1829.

Mackie, R.L., Bennett, B.R., Madden, T.R. (1988). Long period magnetotellutic measurements near the central California coast : a land-locked view of the conductivity structure under the Pacific ocean. Geophys. J. Royal Astr. Soc., 95, 181-194.

Mareschal, M. and Vasseur, G. (1984). Bimodal induction in non-uniform thin sheets : do the present algorithms work for regional studies. J. Geophys., 55, 203-213.

Morse, P. M., Feshback, II. (1953). Methods of theoritical physics. Mc Graw-Hill New York.

Press, S.J. (1989). Bayesian statistics : principle, models ans applications. John Wiley&sons.

Tarantola, A. (1987). Inverse problem theory. Elsevier Science Publishers B. V..

Tarits, P., Jouanne, V., Menvielle, M., Roussignol, M. (1991). Bayesian statistics in geophysics : example of the magnetotelluric 1-d inverse problem. submitted to J. Geophys..

Vasseur, G. and Weidelt, P. (1977). Bimodal electromagnetic induction in non-uniform thin sheets with an application to thr nothern Pyrenean induction anomaly. Geophys. J. R. astr. Soc., 51, 669-690.

Weidelt, P. (1975). Electromagnetic induction in three-dimensional structures. J. Geophys., 41, 85-109.

Markov Random Field Models in Image Remote Sensing

VINCENT GRANVILLE[*]   JEAN-PAUL RASSON[*]

Abstract

During the last few years, Markov Random Field (Mrf) models have already been successfully applied in some applications in image remote sensing in a context of conditional maximum likelihood estimation. Here, in the same context, we propose some original uses of Mrf, especially in image segmentation, noise filtering and discriminant analysis. For instance, we propose a Mrf model on the spectral signatures space, a strongly unified approach to classification and noise filtering as well as a particular model of noise.

Keywords: Supervised classification, segmentation, noise filtering, Markov random field, ICM algorithm

1   Introduction

Since 1970, there has been an increasing interest in digital image processing. In many fields, ranging from medicine, biology, teledetection, chemistry, agronomy, marketing,... image analysis has become a quite common task with the

[*]Département de Mathématique, F.U.N.D.P., Rempart de la Vierge 8, B-5000 Namur, Belgium

introduction on the market of graphic cards and monitors displaying simultaneously up to 16,777,216 colors at a resolution generally above 1024x768 pixels and a price always decreasing. On the other hand, large quantities (e.g. 4 Mgbytes) of fast random access memory (RAM) are now available at low price, and 32 bits processors (i80386/387) are widely spread. These three ingredients are the minimum hardware requirements to allow professional and quite fast *raster* image processing using sophisticated data structures and intensive mathematical computations.

The most common applications requiring non trivial mathematical algorithms in 2-dimensional image analysis are image enhancement (filtering noise, edge enhancement, geometric corrections), segmentation (including edge detection, erosion and dilatation), supervised classification, encoding, compaction, simulation and pattern recognition. We shall be only interested here in the statistical applications based on Bayesian models: we shall describe a general model and algorithm for supervised classification and filtering, with a unified approach for these two problems generally treated differently.

If we note X, Z and N the true unknown image (to be estimated), the observed image and the noise process, then the main problems can be expressed as follows: filtering or classifying an image consists in finding a realization x which maximizes $P(X = x|Z = z)$, that is, by virtue of the Bayesian rule, an x which maximizes $P(Z = z|X = x)P(X = x)$. The difference between both problems will reside in the kind of grid we are working with (d-dimensional spectral grid for classifying and 2-dimensional geometrical grid for filtering) as well as in the kind of Markov random field and noise used for modeling. It is worth noticing that in the classification problem, the noise process is not in fact to be understood as "real noise", but in order to get the same approach for both problems, N can be handled as a noise process. Finally, we shall see on an example how segmentation can improve a classification by including spatial information to the spectral information.

Bayesian methods together with Markov random field models have already been extensively used in the context of image remote sensing (Besag, 1986; Geman and Geman, 1984; Cross and Jain, 1983; Derin, 1986; Derin and Elliott, 1987; Geman, 1988; Lakshmanan and Derin, 1989; Mardia, 1989).

2   Definitions

A digital image is a rectangular grid such that each cell (also called site) is associated to a pixel. A discrete value is assigned to each pixel, e.g the spectral signature. The above grid which is associated to the geometrical coordinates of a pixel will therefore be called the geometrical grid. It is 2-dimensional and contains m sites labelled $1, \ldots, m$. On the other hand, the grid associated to the spectral coordinates is called the spectral grid and it is supposed to be d-dimensional. It contains $M = L^d$ sites labelled $1, \ldots, M$, where L is the number of colors or gray levels (generally $L = 256$).

Whatever the grid we are dealing with, a random variable X_r and a neighboring subset G_r is associated to each site r. The neighboring subsets must satisfy to the two next axioms :

$$\forall r, r \notin G_r$$

$$\forall r, s, (r \in G_s \Rightarrow s \in G_r)$$

Now we can define a Markov random field (Mrf) on either the geometrical or spectral grid. Such a discrete multivariate stochastic process is characterized by the Markov property $P(X_r = x_r | X_s = x_s, s \neq r) = P(X_r = x_r | X_s = x_s, s \in G_r)$. These conditional probabilities are called *local characteristics* and noted $Q_r(.)$ while the joint probability is known to be a Gibbs density by virtue of the Hamersley-Clifford theorem (Besag, 1974).

In the context of image remote sensing, we develop in this paper a new model of noise. Then a classical Mrf on the geometrical grid together with the use of the ICM filtering algorithm allow us to built a post-classification filter based on conditional maximum likelihood inference. The same ideas, but this time applied to the spectral grid, provide a supervised classification algorithm based on kernel density estimation. The computational aspects of the problem are also studied. Finally, a special model of Mrf for image segmentation is exhibited.

3   Mrf and image filtering

The following notations will be used :

- $X = (X_1, \ldots, X_m)$: true image (unknown, to be estimated)

- $Z = (Z_1, \ldots, Z_m)$: observed image (which is a noisy image)

- $N = (N_1, \ldots, N_m)$: noise process

3.1 The model

In order to filter a classified image, we developed a mixture model of noise where the classified image plays the role of Z. As Z contains misclassified and non classified pixels, these pixels can be regarded as noisy pixels. And of course, X stands for the correct (improved) classification. Thus,

If $X_r = N_r$ Then $Z_r = X_r = N_r$

Else either $Z_r = X_r$ with probability $p > 0$, either $Z_r = N_r$ with probability $q = 1 - p$

It is assumed that $\forall r$, X_r and N_r are independent and that $(X_1, \ldots, X_m)$ is a Mrf on the geometrical grid, relatively to a neighboring family $G = \{G_1, \ldots, G_m\}$. The first assumption has already been relaxed on some practical examples. If no information is available on the noise, then the marginals N_r are supposed to be uniform.

3.2 Statistical inference

Our aim is to estimate X by mean of conditional maximum likelihood. Following Besag's approach (Besag, 1986) we shall estimate X using one iteration of his "ICM" algorithm as follows:

1. Take an initial estimate $\hat{x}^{\circ} = z$,

2. Built the estimate $\hat{x}$ as follows:

$$\forall r = 1, \ldots, m, \ \hat{x}_r = \arg \max P(X_r = x_r | Z = z, X_j = \hat{x}_j^{\circ}, j \neq r)$$

Under the assumption that $P(Z = z | X = x) = \prod_{r=1}^{m} P(Z_r = z_r | X_r = x_r)$ (Besag, 1986) we have:

$$P(X_r = x_r | Z = z, X_j = x_j, j \neq r) \propto$$
$$P(Z_r = z_r | X_r = x_r) P(X_r = x_r | X_j = x_j, j \neq r).$$

The second factor of the product is the Mrf local characteristic at site r. The first factor further can be rewritten as follows:

$$P(Z_r = z_r | X_r = x_r) = \begin{cases} p + q\lambda_{z_r} & \text{if } z_r = x_r \\ q\lambda_{z_r} & \text{if } z_r \neq x_r, \end{cases}$$

where

$$\lambda_{z_r} = P(N_r = z_r)$$

Thus, we finally get, with a uniform noise (that is $\lambda_{z_r} = 1/n$) :

$$\hat{x}_r = \begin{cases} z_r & \text{if } Q_r(\bar{c}) < (1 + pn/q)Q_r(z_r) \\ \bar{c} & \text{else} \end{cases}$$

where

- n = number of possible values for the noise marginal

- $\bar{c}$ is the mode of the local characteristic $Q_r()$

The ICM algorithm is known to converge very quickly to a local maximum of $P(X = x | Z = z)$, after 5 or 6 cycles. Here, we stop after the first cycle because further cycles can degrade the image: although more cycles would provide an estimate closer to the true *theoretical* image, in practice the second and third cycles may increase the smoothing and clean very little spots that are in fact not noisy. The reason is probably the fact that practical images do not follow exactly the simple Mrf models described here.

3.3 Application : a post classification filter

When the observed image Z is a classification as described above, the simplest local characteristics that can be used on the geometrical grid are

$$Q_r(c) \propto \exp(\tau \sum_{s \in G_r} \chi_{\{x_s = c\}})$$

Then the filtering rule simplifies in

$$\hat{x}_r = \arg\max(\alpha \cdot \chi_{\{z_r = c\}} + \beta \sum_{s \in G_r} \chi_{\{z_s = c\}})$$

where

- $\alpha, \beta > 0$

- $\alpha + \beta = 1$

- $\alpha\tau = \beta \log(1 + pn/q)$

- $\tau = 1.5$ (Besag, 1986)

The parameters α, β are nothing else than a reparametrization of p and τ. But their interpretation is straightfoward : they are the weights attributed to the two first windows that constitue the neighborhood of any pixel. Thus a high value of β will provide a powerfull filter and conversely. Note that the effect of one cycle of ICM with a high value for β is quite the same as the effect of two cycles with a low value of β or also one cycle with a low value of β and a larger size of the neighboring subset. Of course, the first approach is more efficient. Although estimation of α and β can be derived e.g by mean of the coding pattern (Besag, 1974), we have always applied this algorithm successfully with the same fixed empirical values. In fact, as stated in the next theorem, there are only a finite number of classes of parameters, and among these classes, many produce a similar filtered image.

Theorem 3.1 *The parameters α and β partition the simplex $\alpha + \beta = 1, \alpha, \beta > 0$ in a finite number of convex classes as follows : two couples of parameters belong to a same class iff they produce the same filtered image*

$\square$

3.4   CPU time

The filtering algorithm relics simply on counting techniques and can be written only with integers. Since one cycle of ICM is used, it is much more faster than a simulated annealing algorithm.

4   Mrf and image classification

The following notations will be used :

- $X = (X_1, \ldots, X_M)$: classified image (to be estimated)

- $Z = (Z_1, \ldots, Z_M)$: initial image (training sets)

- $N = (N_1, \ldots, N_M)$: noise process

4.1 Theoretical model and inference

We are now working on the spectral grid, in a context of supervised classification. On each site r, we observe a value z_r, realization of Z_r. If $z_r = 0$, then this means that no training set point is located at site r. If $z_r \neq 0$, then z_r is the class of the training set point located at site r. Thus it is assumed that two training set points with identical spectral signatures can not belong to different classes.

The model used for classification is the same as the one used in filtering. The only difference is the fact that the noise is a constant random variable : $\forall r, N_r = 0$. In this context, a noisy site (on the spectral grid) is an unclassified site. As the classes are labelled $1, \ldots, n$, if $z_r \neq 0$ then $x_r = z_r$. Therefore, we get that

$$P(Z_r = z_r | X_r = x_r) = \begin{cases} 1 & \text{if } z_r = x_r = 0 \\ p & \text{if } z_r = x_r, x_r \neq 0 \\ q & \text{if } z_r = 0, x_r \neq 0 \\ 0 & \text{otherwise,} \end{cases}$$

One of the simplest local characteristics for unordered colors that correspond to unordered classes, and preventing sites being unclassified (i.e. taking value 0), is as follows :

$$Q_r(c) \propto \begin{cases} \exp\left(\tau \sum_{s \in G_r} \chi_{\{x_s = c\}}\right) & c \neq 0 \\ \exp\left(-\tau \sum_{s \in G_r} \chi_{\{x_s = c\}}\right) & c = 0, \end{cases}$$

On the spectral grid, the neighborhood of any site will be the hypercube of the length h. The statistical inference remains identical : a single cycle of ICM is used to get an approximate estimator $\hat{x}$ of the classification. Then, provided that $q \exp \tau > 1$, the classification rule for the site r becomes:

- If there is a point from one training set at site r (i.e. $z_r \neq 0$) then assign site r to its associated class (i.e. $\hat{x}_r = z_r$);

- If there is no point from training sets at site r then

- if there is also no point from training sets in the neighborhood G_r then let site r unclassified, i.e. set $\hat{x}_r = 0$.
- in other case classify to the majority in the neighborhood G_r, i.e. set $\hat{x}_r = \arg\max \sum_{s \in G_r} \chi_{\{z_s = c\}}$

In particular, the rate of well-classified points is 100% in training sets. If after one loop of ICM algorithm some site remains unclassified we continue by applying ICM again until all points are classified. However, it is not hard to see that in the latter case sites will be classified according to the 'nearest neighbor' rule and thus it is not necessary to call ICM several times since the nearest neighbor classification can be done separately in the first call of ICM.

In practice, we do not have to classify all grid points but only the portion occupied by spectral signatures of the image. Finally we remark that our Mrf estimation technique has some similarities with the so-called kernel density estimation (Silverman, 1986).

4.2 CPU time

The more efficient algorithm to classify an image relies on range searching techniques used in data base management. If a counting version of the range seaching algorithm is used to count the training set points falling into the hypercube, then the CPU time is output-insentive, i.e it does not depend on the size of the hypercube. In that case, we can expect $O(L \log^{d-1} L)$ for both storage and preprocessing, and $O(\log^{d-2} L \log \log L)$ CPU time to classify one site of the grid, where L is the total number of points in the training sets. For further explanations, the reader is referred to Overmars (1985) and Granville et all. (1992).

In this paper, we develop a naive but powerful and easy-to-implement procedure, as an alternative to the quite sophisticated range searching algorithm. However, our procedure is output-sensitive and should be used with small sizes of the hypercube; it is specially intended to FORTRAN programers. Now we give the details of the data stucture (the associated search is straitghfoward).

The training set points are accessed via a linear array $BASE$, specially organized. Two $n \times 256 \times 256$ arrays called $INDEX$ and NPT are used. $INDEX(i, x, y)$ is a pointer to the first training set point belonging to class i and with two first spectral coordinates x and y. $NPT(i, x, y)$ is the number of such points in all the training sets. These points are finally numbered $BASE(INDEX(i, x, y)), BASE(INDEX(i, x, y) + 1), \ldots, BASE(INDEX(i, x, y) + NPT(i, x, y) - 1)$, and ordered with respect to their third spectral coordinate, to allow a dichotomic search. With this structure, the search is limited to a small part of $INDEX$.

This simple idea has also successfully been applied to compute very quickly the max min of the maxidistances in the training sets, using hypercubes of growing size until all hypercubes centered on one training set point contain another training set point. The final hypercube was then used as an estimator of G_r.

5   Mrf and image segmentation

The following notations will be used :

- $X = (X_1, \ldots, X_m)$: true scene

- $Y = (Y_1, \ldots, Y_K)$: segmentation into K segments (to be estimated)

In this section we propose a model for image segmentation. By segmentation, we mean a partition of the image in some rather homogeneous spots called segments. We do not speak here about statistical inference for at this time, we have not used the model to perform a segmentation $\hat{y}$ that maximizes $P(Y = y|X = x)$, e.g. by mean of a simulated annealing algorithm. Our purpose is only to exhibit an original Mrf on the geometrical grid.

5.1   The model

In its simplest form, the model, which already appeared in the litterature (Lee and Crawford, 1989) is given by

$$P(Y = y|X = x) \propto P(Y = y) \prod_{k=1, K} P(X_r = x_r, r \in S_k | Y = y)$$

where S_k stands for segment k.

It is costumary to assume conditional independence inside a segment, that is $P(X_r = x_r, r \in S_k | Y = y) = \prod_{r \in S_k} P(X_r = x_r | Y = y)$. On the other hand,

$P(Y = y)$ has already been modelized by a Mrf over the geometrical grid. The most interesting thing is the fact that conditionally to the segmentation, the true scene follows what we have called a K-segmented Mrf on the geometrical grid : this means that the neighborhood of any pixel is contituted by all the other pixels belonging to the same segment. The exact definition is given just below.

A *segmented Mrf* relative to a neighboring family $G = \{G_1, \ldots, G_m\}$ is a Mrf satisfying the following axiom :

$$\forall r, s, t, (r \in G_s, s \in G_t \text{ and } r \neq t) \Rightarrow r \in G_t$$

Such a Mrf is said to be *K-segmented* if among $G_1 \cup \{1\}, \ldots, G_m \cup \{m\}$, there are only K distinct subsets $(1 \leq K \leq m)$.

Theorem 5.1 *A Mrf with a joint density $P(X)$ is K-segmented iff there is a partition $\{S_1, \ldots, S_K\}$ of the associated grid such that $P(X) = \prod_{k=1,K} P(X_r, r \in S_k)$*

Proof:
If $P(X)$ factorizes, then it is easy to see that $P(X_i|X_j, j \neq i) = P(X_i|X_j, j \in S_{l_i})$, where S_{l_i} is the segment containing site i.

It is not so easy to prove the reciprocal. Let X be a K-segmented MRF into K segments $S_1, \cdots, S_K$ and $\mathcal{U} = S_1 \cup \cdots \cup S_K$. Let $S'_1 = \mathcal{U} \setminus S_1$. We shall prove that $P(X)$ factorizes in $P(X) = P(X_i, i \in S_1)P(X_i, i \in S'_1)$. Then it suffices to factorize recursively the second factor to get the thesis. W.l.o.g, we shall assume that $S_1 = \{1, \cdots, s\}$ and we shall use the following notations:

$$\begin{aligned}
\pi &= P(X) \\
\pi' &= P(X_i, i \in S_1) \\
\pi_{i_1, \cdots, i_p} &= P(X_i, i \neq i_1, \cdots, i \neq i_p) \\
\pi'_{i_1, \cdots, i_p} &= P(X_i, i \neq i_1, \cdots, i \neq i_p, i \in S_1)
\end{aligned}$$

We shall give the demonstration for $s = 3$. The generalization to an arbitrary value of s is only a problem of notations. If $s = 3$, we get successively

$$\pi_1 = \pi\pi'_1/\pi', \quad \pi_2 = \pi\pi'_2/\pi', \quad \pi_3 = \pi\pi'_3/\pi' \tag{1}$$
$$\pi_1 = \pi_3\pi'_1/\pi'_3, \quad \pi_2 = \pi_3\pi'_2/\pi'_3$$

$$\pi_{13} = \pi_3 \pi'_{13}/\pi'_3, \quad \pi_{23} = \pi_3 \pi'_{23}/\pi'_3 \qquad (2)$$
$$\pi_{13} = \pi_{23} \pi'_{13}/\pi'_{23}$$

$$\pi_{123} = \pi_{23} \pi'_{123}/\pi'_{23} \qquad (3)$$

But $\pi'_{123} = 1$ and $\pi_{123} = P(X_i, i \in S'_1)$. Thus, we have

$$
\begin{aligned}
\pi_{123} &= \pi_{23}/\pi'_{23} \quad \text{by } 3 \\
&= \pi_3/\pi'_3 \quad \text{by } 2 \\
&= \pi/\pi' \quad \text{by } 1
\end{aligned}
$$

Therefore, $P(X_i, i \in S'_1) = P(X)/P(X_i, i \in S_1)$.

$\square$

6   A real life application

In this section we give an illustration of the filtering and classification algorithm on a 512×512 image with known true scene. Therefore, our error rates directly computed on the true scene X and expressed in terms of confusion matrix are more reliable than those produced by leaving-one-out or resubstitution methods. The segmentation used here was provided by IGN (France). Here, the segmentation-based classification is simply a classification where each segment S_k is assigned globally to a group using the majority rule on the point-by- point classification of S_k. The true scene, the point-by-point classification, the filtered point-by-point classification and the segmentation-based classification are respectively noted $X, Z, \hat{X}$ and Y. We compare our results with a quadratic multinormal classification U computed with IMSL.

The training set contains about 16,000 points into 7 classes numbered $1, \cdots, 7$. Class 0 contains all the points that have not been classified after the first cycle of ICM. The size of the neighboring hypercube in the spectral grid has been made dependent on the class number and estimated (for each class) by pseudo maximum likelihood: this improves the classification (but not the global error rate), especially for small classes.

Figure 1: *True scene and classification of a 3-dimensional SPOT image by use of a "filtering" algorithm*

X \ U	0	1	2	3	4	5	6	7
1	0	28	26	3	2	4	21	11
2	0	5473	49259	1480	488	1078	1660	242
3	0	223	417	6225	135	1055	413	54
4	0	4909	485	653	27964	874	4291	10235
5	0	1997	1090	18386	968	15983	3718	431
6	0	737	435	285	135	222	891	205
7	0	40	0	1	145	6	20	515

Table 1: *Confusion matrix between the true scene X and a quadratic multi-normal classification U*

Z X	0	1	2	3	4	5	6	7
1	0	47	13	0	6	2	27	0
2	62	280	50305	424	981	3640	3946	42
3	87	1	201	5177	125	2708	222	1
4	282	27	379	180	41732	2264	3237	1310
5	215	25	814	4152	1639	33607	2087	34
6	4	5	306	31	430	486	1624	24
7	5	0	0	0	299	5	13	405

Table 2: *Confusion matrix between the true scene X and our point-by-point classification Z*

$\hat{X}$ X	0	1	2	3	4	5	6	7
1	0	45	13	0	7	1	29	0
2	0	73	52124	147	937	3513	2881	5
3	25	0	204	5144	117	2855	177	0
4	33	2	420	161	43665	2357	2517	256
5	1	10	750	2840	1581	35846	1538	7
6	0	0	259	13	370	528	1736	4
7	0	0	0	0	342	6	2	377

Table 3: *Confusion matrix between the true scene X and our filtered classification $\hat{X}$*

Y / X	0	1	2	3	4	5	6	7
1	0	49	0	0	0	0	46	0
2	0	0	57739	0	0	1087	854	0
3	0	0	0	5189	0	3204	129	0
4	0	0	0	0	47737	868	806	0
5	0	0	0	2013	0	40172	388	0
6	0	0	175	0	307	423	2005	0
7	0	0	0	0	404	0	0	323

Table 4: *Confusion matrix between the true scene X and our segmentation-based classification Y*

References

[1] BESAG J. (1974): *Spatial interaction and the statistical analysis of lattice systems*, JRSS B, 36, 192–226.

[2] BESAG J. (1986): *On the statistical analysis of dirty pictures*, JRSS B, 48, 259–302.

[3] CHAZELLE (1988): *A functional approach to data structures and its use in multidimensional searching*, SIAM J. Comp. 17, 427–462.

[4] CROSS G.R., JAIN A.K. (1983): *Markov random field texture models*, IEEE Trans. PAMI 5, 25–39.

[5] DERIN H. (1986): *Segmentation of textured images using Gibbs random fields*, Comp. Vision, Graphics and Image Processing, 35, 72–98.

[6] DERIN H., ELLIOTT H. (1987): *Modeling and segmentation of noisy and textured images using Gibbs random fields*, IEEE Trans. PAMI 9, 39–55.

[7] GEMAN D. (1988): *Random fields and inverse problems in imaging*, Lectures Notes in Math. 1427, 117–196.

[8] GEMAN S., GEMAN D. (1984): *Stochastic relaxation, Gibbs distributions and the Bayesian restoration of images*, IEEE Trans. PAMI 6, 721–741.

[9] GRANVILLE V., KRIVANEK M., RASSON J.P. (1992): *Clustering, classification and image segmentation on the grid,* to appear in Comp. Stat. Data Analysis

[10] GRANVILLE V., RASSON J.P. (1992): *A new modelisation of noise in image remote sensing,* to appear in Stat. Probability Letters

[11] GRANVILLE V., RASSON J.-P. (1992): *A Bayesian filter for a mixture model of noise in image remote sensing,* to appear in Comp. Stat. Data Analysis.

[12] GUYON X. (1985): *Champs stationnaires sur Z^2: modèles, statistique et simulations,* Tech rep. Université Paris I.

[13] LAKSHMANAN S., DERIN H. (1989): *Simultaneous parameter estimation and segmentation of Gibbs random fields using simulated annealing,* IEEE Trans. PAMI 8, 799–813.

[14] LEE S., CRAWFORD M.M. (1989): *Statistically based unsupervised hierarchical image segmentation algorithm with a blurring corrector,* Proc. of the 12-th Canadian Symp. on Remote Sensing 2, 630–633.

[15] MARDIA K.V. (1989): *Markov models and Bayesian methods in image analysis,* JAP 16, 125–130.

[16] MONGA O. (1990): *Image segmentation: state of the art,* tutorial for PIXIM'89 conference; also available as INRIA Tech. rep. No. 1216, Rocquencourt (in French).

[17] OVERMARS M.H. (1985): *Range searching on the grid,* Proc. workshop on graphtheoretic concepts in computer science, Trauner Verlag, 295–305.

[18] SILVERMAN B.W. (1986): *Density estimation for statistics and data analysis.* Chapman and Hall, NY.

Minimax Linewise Algorithm for Image Reconstruction

A.P. Korostelev and A.B. Tsybakov

A.P. Korostelev
Institute for System Analysis
Prosp. 60 – letija Oktjabrja 9
117312 Moscow, Russia

A.B. Tsybakov
Institute de Statistique
Université Catholique de Louvain
B-1348 Louvain-la-Neuve, Belgium

ABSTRACT

We study the problem of estimating the edges in noisy images by linewise procedures. We show that the straightforward estimation method (naïve linewise procedure) does not attain the asymptotically minimax rate of accuracy as the number of observations tends to ∞. We propose the modified linewise procedure which has the asymptotically minimax rate.

AMS 1980 Subject Classification 62G05

Key words and phrases: image estimation, edge estimation, minimax rate of convergence, linewise algorithm, nonparametric regression, change-point problem.

MINIMAX LINEWISE ALGORITHM FOR IMAGE RECONSTRUCTION

1. Introduction

There exists an extensive literature on image analysis (see e.g. the books of Pratt (1978), Rosenfeld and Kak (1982), Marr (1982), Blake and Zisserman (1987)). One of the important problems in image analysis is reconstruction of pictures from noisy data, or image estimation. There are two special features related to this problem. First, the data arrays are two-dimensional (or multidimensional). Second, the image is usually composed of several regions with rather sharp edges. Within each region the image preserves a certain degree of uniformity while on the boundaries between the regions it has considerable changes. Therefore, it is important to find the boundaries (edges) of such regions. This leads to edge detection or edge estimation problems.

A large variety of methods has been proposed for solving the image and edge estimation problems in different contexts. An overview of them can be found in the above mentioned books. The most popular methods are based on certain parametric image models with large number of parameters and on Bayesian approach to estimation (Haralick (1980), Geman and Geman (1984), Ripley (1988) among others). The methods of edge detection, however, often have the local character and do not assume any underlying parametric model (Marr (1982), Nagao and Matsuyama (1979), Huang and Tseng (1988)). Another nonparametric approach to the construction of image estimators is related to penalizing and regularization techniques (Titterington (1985), Torre and Poggio (1986), Shiau, Wahba and Johnson (1986), Mumford and Shah (1989), Girard (1990)).

In Tsybakov (1989), Korostelev (1991), Korostelev and Tsybakov (1992) the minimax approach to image and edge estimation was developed which makes possible to compare different image and edge estimators on the common scale. This approach is similar to the minimax theory of nonparametric regression and density function estimation (Ibragimov and Khasminskii (1979), Bretagnolle and Huber (1979), Stone (1980, 1982)).

It has been shown in Korostelev and Tsybakov (1992) that the best possible rates of convergence (minimax rates) in various image estimation problems are attained by some estimators which are, however, complicated from the computational viewpoint. On the other hand, there exist many simple data processing algorithms in image analysis which do apply in practice. We study here one of them: the linewise algorithm, and show that it does not have the minimax rate of convergence in its naïve form. Next we propose a modified linewise algorithm which attains the best possible (minimax) rate. The study of linewise procedures reveals in the most obvious form the link between image estimation, change-point problem and nonparametric regression.

2. Edge Estimation Problem

Consider a simple two-dimensional image model. Assume that G is a set that belongs to the unit square $K = [0, 1] \times [0, 1]$. Let n_1 be an integer.

Let the sample of $n = n_1^2$ observations is given,

$$(2.1) \qquad Y_{ij} = I(X_{ij} \in G) + \xi_{ij}, \qquad i, j = 1, \ldots, n_1,$$

where $X_{ij} \in K$ are the design points, ξ_{ij} are i.i.d. $(0, \sigma^2)$-Gaussian random errors and $I(.)$ denotes the indicator function.

We assume that G is of the form

$$G = \{x = (x_1, x_2) \in K : \quad 0 \le x_1 \le 1, \quad 0 \le x_2 \le g(x_1)\}$$

where $g \in \Sigma(\gamma, L, h), \gamma > 1, L > 0, 0 < h < 1/2$, and $\Sigma(\gamma, L, h)$ denotes the class of functions $g(x_1), h < g(x_1) < 1 - h$, defined on the interval $[0, 1]$ and having the Hölder kth derivative:

$$| g^{(k)}(x_1) - g^{(k)}(x_1') | \le L | x_1 - x_1' |^{\gamma - k}, x_1, x_1' \in [0, 1],$$

where k is the maximal integer such that $k < \gamma$. Denote by $\mathcal{G}(\gamma, L, h)$ the class of all domains G satisfying these conditions.

Define the design points

$$(2.2) \qquad X_{ij} = (x_{ij}^{(1)}, x_{ij}^{(2)}) = (i/n_1, j/n_1 - \eta_i), \quad i, j = 1, \ldots, n_1,$$

where η_i are independent random variables uniformly distributed in $[0, 1/n_1]$. The design (2.2) is composed of random shifts of the regular grid; these shifts are independent for each line $x_1 = i/n_1, \quad i = 1, \ldots, n_1$. We assume everywhere that η_i's are independent of $\xi_{ij}, i, j = 1, \ldots, n_1$.

A set $G \in \mathcal{G}(\gamma, L, h)$ is called *image*. An estimator $\hat{G}_n$ of image G is assumed to be a closed subset of K measurable with respect to observations (X_{ij}, Y_{ij}), $i, j = 1, \ldots, n_1$. For an arbitrary estimator $\hat{G}_n$ introduce the maximal risk

$$r_n\left(\hat{G}_n, \psi_n\right) = \sup_{G \in \mathcal{G}(\gamma, L, h)} E_G\left[\psi_n^{-1} d(G, \hat{G}_n)\right]$$

where E_G is the expectation with respect to the distribution of observations (X_{ij}, Y_{ij}), $i, j = 1, \ldots, n_1$, ψ_n is a normalizing factor, and $d(G, \hat{G}_n)$ is the Lebesgue measure of symetric difference $G \triangle \hat{G}_n$,

$$d(G, \hat{G}_n) = \text{mes } (G \triangle \hat{G}_n).$$

Note that the boundary of $\hat{G}_n$ is not necessarily one-valued function of x_1.

The problem of estimation of G (or, equivalently, of the function g) from observations (2.1.) is called *edge estimation problem*.

The positive sequence ψ_n is called *minimax rate of convergence* and the estimator G_n^* is called *optimal estimator* if

(2.3) $$\lim_{n \to \infty} \sup r_n(G_n^*, \psi_n) \le c_1,$$

and for any estimator $\hat{G}_n$

(2.4) $$\lim_{n \to \infty} \inf r_n(\hat{G}_n, \psi_n) \ge c_2$$

where c_1, c_2 are positive constants.

As proved by Korostelev and Tsybakov (1992) for slightly different designs the minimax rate of convergence in this problem is equal to $n^{-\gamma/(\gamma+1)}$. In Korostelev and Tsybakov (1992) the optimal estimator G_n^* satisfying (2.3) with $\psi_n = n^{-\gamma/(\gamma+1)}$ was proposed based on combination of piecewise-polynomial approximation with maximum likelihood estimator (MLE).

Though edge estimation problem considered here is not formally covered by the result of Korostelev and Tsybakov (1992), nevertheless, it is a routine technique to extend it to the design (2.2). The aim of this paper is to construct another optimal estimator, based on solutions of change-point problems for each line $x_1 = i/n_1$, $i = 1, \ldots, n_1$.

3. Naïve Linewise Processing

Note that for each line $x_1 = i/n_1$ we have an independent change-point problem with the change-point parameter $\theta_i = g(i/n_1)$, $i = 1, \ldots, n_1$ (the notation θ_i should be provided with index n which is omitted for the sake of brevity). It is necessary to emphasize that we study the posterior change-point problem as in Ibragimov and Khasminskii (1979), Sec.7.2, but not as a problem of sequential analysis (cf. Siegmund (1985)).

Denote by θ_{in}^* the MLE for θ_i obtained from observations

$$((X_{i1}, Y_{i1}), \ldots, (X_{in_1}, Y_{in_1})), \quad i = 1, \ldots, n_1.$$

The MLE's θ_{in}^* are independent for different i. We describe now some asymptotic properties of the MLE in assumption that n tends to infinity.

For the Gaussian errors in (2.1) the MLE θ_{in}^* coincides with the least squares estimator (LSE) which is the solution of the following minimization problem:

$$\min_{\theta} \sum_{j=1}^{n_1} \left[Y_{ij} - I\left(X_{ij} = \left(x_{ij}^{(1)}, x_{ij}^{(2)} \right) : x_{ij}^{(2)} \leq \theta \right) \right]^2$$

or equivalently,

(3.1) $$\max_{\theta} J_i(\theta),$$

$$J_i(\theta) = \sum_{j=1}^{n_1} I\left(X_{ij} : x_{ij}^{(2)} \leq \theta \right) Y_{ij} - (1/2) \operatorname{card} \left\{ i : x_{ij}^{(2)} \leq \theta \right\}.$$

Note that $J_i(\theta)$ are right-continuous piecewise constant functions having jumps at points $x_{ij}^{(2)}$. Due to the Gaussian distribution of errors in (2.1) there exists for each i a unique $\hat{j} = \hat{j}(i)$ such that any solution of (3.1) belongs to the interval $\left[x_{i,\hat{j}-1}^{(2)}, x_{i,\hat{j}}^{(2)} \right)$. We assume that the maximization in (3.1) is over $\theta \in [x_{i1}^{(2)}, x_{in_1}^{(2)})$, and thus possible boundary effects are neglected. It is convenient to define the MLE as the mean value of the extreme solutions; this can be written as

$$(i/n_1, \theta_{in}^*) = \left(X_{i,\hat{j}-1} + X_{i,\hat{j}} \right) / 2, 1 < \hat{j} \leq n_1.$$

In the following we assume that $\sigma^2 = 1$ without loss of generality. Define a two-sided random walk Z_k:

$$Z_0 = 0, \quad Z_k = W_k - (1/2) \mid k \mid, \quad k = \pm 1, \pm 2, \ldots,$$

where

$$W_k = \begin{cases} \Sigma_{i=1}^{k} \bar{\xi}_i \text{ if } k > 0, \\ \Sigma_{i=k}^{-1} \bar{\xi}_i \text{ if } k < 0, \end{cases}$$

$\bar{\xi}_i$ are independent standard $(0,1)$-Gaussian random variables. Let $\hat{k}$ be the maximizer for Z_k, i.e. $Z_{\hat{k}} > Z_k$ for $k \neq \hat{k}$. Note that the paths $Z_k \to -\infty$ with probability 1 as $|k| \to \infty$. Hence, the point $\hat{k}$ is unique and finite almost surely. Introduce the probabilities

$$\pi_k = P(\hat{k} = k), \quad k = 0, \pm 1, \ldots,$$

where P is probability associated with Z_k.

3.1. Lemma. *The probabilities π_k are positive, symmetric, i.e. $\pi_k = \pi_{-k}$, and $\pi_k \leq \exp(-\lambda \mid k \mid)$ with some $\lambda > 0$ and $|k|$ large enough.*

Proof. Using the inequality $1 - \Phi(x) \leq (2/\pi)^{1/2} \exp(-x^2/2)$, $x > 1$, which holds for standard Gaussian distribution function $\Phi(x)$, one gets

$$\pi_k \leq P(\hat{k} \geq k) \leq P\left[\sup_{k_1 \geq k} Z_{k_1} \geq 0\right] \leq \sum_{k_1 \geq k} P\left[k_1^{-1/2} W_{k_1} \geq k_1^{1/2}/2\right] =$$

$$= \sum_{k_1 \geq k} \left(1 - \Phi(k_1^{1/2}/2)\right) \leq 2\pi^{-1/2} \sum_{k_1 \geq k} \exp(-k_1/8) \leq \exp(-\lambda k)$$

for some positivie λ and k large enough. The same argument is used for $k < 0$. ∎

Define the symmetric probability density

$$\pi(x) = \pi_k \text{ if } k - 0.5 < x < k + 0.5, \quad k = 0, \pm 1, \ldots,$$

3.2. Lemma. *There exists a positive constant λ_1 such that for n_1 large the following inequalities hold uniformly in $G \in \mathcal{G}(\gamma, L, h)$*

$$\sup_{-\infty < x < \infty} |P_G(n_1(\theta_{in}^* - \theta_i) \leq x) - \int_{-\infty}^{x} \pi(y)dy| \leq \exp(-\lambda_1 n_1),$$

and

$$|E_G(n_1(\theta_{in}^* - \theta_i))| \le \exp(-\lambda_1 n_1/2).$$

Proof. Let an integer $j_0 = j_0(\theta_i)$ be such that $x_{i,j_0-1}^{(2)} \le \theta_i < x_{ij_0}^{(2)}$, and let $\overline{\theta}_i$ be the midpoint of this interval, i.e.

$$\overline{\theta}_i = (x_{i,j_0-1}^{(2)} + x_{ij_0}^{(2)})/2.$$

We have

$$n_1(\theta_{in}^* - \theta_i) = n_1(\theta_{in}^* - \overline{\theta}_i) + n_1(\overline{\theta}_i - \theta_i),$$

where $n_1(\overline{\theta}_i - \theta_i)$ is uniformly distributed in the interval $(-0.5, 0.5)$. Note that the random integer j_0 depends only on the design $\mathcal{X}_i = (x_{i1}^{(2)}, \ldots, x_{in_1}^{(2)})$ and does not depend on the random variables ξ_{ij}.

Set $h_0 = h/2$, $k_n = [n_1 h_0]$, and define the random event

$$\mathcal{A} = \{\mathcal{X}_i : n_1 h_0 \le j_0 \le n_1(1 - h_0) + 2\}.$$

Let us show that for $n > 3/h_0$ and for any $\mathcal{X}_i \in \mathcal{A}$ we have

$$(3.2) \qquad |P_G(\hat{j} = k + j_0 | \mathcal{X}_i) - \pi_k| \le 2\exp(-\lambda(n_1 h_0 - 3))(a.s.), |k| \le k_n.$$

In fact,

$$P_G(\hat{j} = k + j_0 | \mathcal{X}_i) \le \pi_k + P(\hat{k} + j_0 \notin \{2, \ldots, n_1\}).$$

By Lemma 3.1

$$P(\hat{k}+j_0 \notin \{2, \ldots, n_1\}) \le \sum_{|k| \ge n_1 h_0 - 2} \pi_k \le 2 \sum_{k \ge n_1 h_0 - 2} \exp(-\lambda k) \le 2\exp(-\lambda(n_1 h_0 - 3)).$$

This gives the estimation for the probability in (3.2) from above. The estimation from below is quite similar.

For any integer k, $|k| \le k_n$, and for any $x \in (-1/2, 1/2)$, we have
$P_G(n_1(\theta_{in}^* - \theta_i) \in (k - 1/2, k + x)) =$

$$= P_G(n_1(\theta_{in}^* - \overline{\theta}_i) = k, n_1(\overline{\theta}_i - \theta_i) \in (-1/2, x)) =$$
$$= P_G(\hat{j} - j_0 = k, n_1(\overline{\theta}_i - \theta_i) \in (-1/2, x)).$$

Applying (3.2), one has

$$|P_G\left(n(\theta_{in}^* - \theta_i) \in (k - 1/2, k + x)\right) - \int_{k-1/2}^{k+x} \pi(y)dy| \le 2\exp(-\lambda(n_1 h_0 - 3)).$$

Hence for any $x, x \le k_n + 1/2$,

$$|P_G(n_1(\theta_{in}^* - \theta_i) \le x) - \int_{-\infty}^{x} \pi(y)dy| \le$$

$$\le \sum_{k=-k_n}^{[x-1/2]} |P_G(n_1(\theta_{in}^* - \theta_i) \in (k - 1/2, k + 1/2)) - \int_{k-1/2}^{k+1/2} \pi(y)dy| +$$

$$+ |P_G(n_1(\theta_{in}^* - \theta_i) \in (x - 1/2] + 1/2, x)) - \int_{[x-1/2]+1/2}^{x} \pi(y)dy| +$$

$$+ P_G(n_1(\theta_{in}^* - \theta_i) \le -k_n - 1/2) + \int_{-\infty}^{-k_n-1/2} \pi(y)dy \le$$

$$\le 2(2k_n + 1)\exp(-\lambda(n_1 h_0 - 3)) + P_G(\hat{j} - j_0 < -k_n) + \sum_{k \le -k_n} \pi_k.$$

Due to Lemma 3.1 and (3.2) the last two summands decrease exponentially as $n_1 \to \infty$. this proves the first inequality of Lemma 3.2.

Applying this inequality and using the symmetry of $\pi(x)$ and the obvious property $|n_1(\theta_{in}^* - \bar{\theta}_i))| < n_1$ we have that uniformly in $\theta_i \in (h, 1-h)$ the inequality holds

$$|E_G(n_1(\theta_{in}^* - \theta_i))| \le \int_{-\infty}^{\infty} x\pi(x)dx + n_1 \exp(-\lambda_1 n_1)$$

$$= n_1 \exp(-\lambda_1 n_1) \le \exp(-\lambda_1 n_1/2)$$

for n_1 large. ∎

3.3. Corollary. *According to Lemma 3.2 for the MLE's θ_{in}^* the stochastic expansion holds*

(3.3) $$\theta_{in}^* = \theta_i + b_n(\theta_i) + n_1^{-1}\varepsilon_{in}, \quad i = 1, \dots, n_1,$$

where the random errors ε_{in} are zero-mean, independent for different i, and have variances bounded uniformly in i and $n \ge 1$; for each value of i $b_n(\theta_i)$ denotes a bias term satisfying for n large

(3.4) $$|b_n(\theta_i)| \le \exp(-\lambda_2 n_1)$$

with some positive λ_2 uniformly in $\theta_i \in (h, 1 - h)$.

We mean by naïve linewise processing the following two-step procedure. The first step consists of calculation θ_{in}^* for all the lines $x_1 = i/n_1$. At the second step some smoothing algorithm is applied to θ_{in}^*'s which are regarded now as new observations in the nonparametric regression

$$(3.5) \qquad \theta_{in}^* \approx g(i/n_1) + \epsilon_{in}/n_1, \quad i = 1, \ldots, n_1$$

Surely, the bias terms $b_n(\theta_i)$ in (3.3) are negligible as follows from (3.4). If we have estimated the regression function g from observations (3.5) in L_1−norm, this would lead to the same accuracy for the edge estimation in d-metric. It is well-known (see Ibragimov and Khasminskii (1979), Stone (1980, 1982), Härdle (1990)) that the minimax rate of convergence in L_1−norm in nonparametric regression problem (3.5) equals $\psi_n = (\sigma^2/n_1)^{\gamma/(2\gamma+1)}$ where $\sigma^2 = n_1^{-2} = n^{-1}$ is the variance of one observation in (3.5). Hence the best rate of convergence available from observations (3.5) is equal to

$$(3.6) \qquad \psi_n = n^{-3\gamma/(4\gamma+2)}$$

Two conclusions follow from this. First, the rate of convergence obtained by the naïve linewise processing is not the best possible. Indeed, inequality $\gamma/(\gamma+1) > 3\gamma/(4\gamma + 2)$ holds for $\gamma > 1$. Second, the rate of convergence achieved in (3.6) cannot be improved uniformly in $g \in \Sigma(\gamma, L, h)$ if we use only the MLE θ_{in}^* at the second step of the procedure.

Now we concentrate on the next problem: whether it is possible to modify this linewise processing in such a way that the resulting procedure would achieve the minimax rate of convergence? The answer is positive. Certainly, for such a procedure we must have not only the values θ_{in}^* but also some additional information for each line $x_1 = i/n_1, \quad i = 1, \ldots, n_1$.

4. Modified Linewise Procedure

The idea of modification is rather simple. Suppose that at each point $(i/n_1, \theta_{in}^*)$ we are able to get a new observation Y_i^* which is independent both of design (2.2) and of observations (2.1). Then we can try the LSE obtained from the new observations (θ_{in}^*, Y_i^*), $i = 1, \ldots, n_1$. All we have to do is to "extract" an observation from the data which would have the same properties as Y_i^* and which would not spoil nice properties of the new "design points" θ_{in}^*.

In addition suppose that $n_1 = n^{1/2}$ is even. Split each sample $((X_{i1}, Y_{i1}), \ldots, (X_{i,n_1}, Y_{i,n_1}))$, $i = 1, \ldots, n_1$, into odd and even subsamples. Denote by θ_{in}^{odd} and $\theta_{in}^{\text{even}}$ the MLE's obtained respectively from the odd and even subsamples, i.e. from

$$((X_{i1}, Y_{i1}), (X_{i3}, Y_{i3}), \ldots, (X_{i,n_1-1}, Y_{i,n_1-1}))$$

and

$$((X_{i2}, Y_{i2}), (X_{i4}, Y_{i4}), \ldots, (X_{i,n_1}, Y_{i,n_1})).$$

By definition the MLE θ_{in}^{odd} is such that

$$(4.1) \qquad (i/n_1, \theta_{in}^{\text{odd}}) = (X_{i,\hat{j}-1} + X_{i,\hat{j}+1})/2 = X_{i,\hat{j}},$$

for some $\hat{j} = \hat{j}(i), 1 < \hat{j}(i) < n_1, \hat{j}(i)$ even (cf. Section 3).

This is the only point where the technical convenience of the design (2.2) appears: the MLE strictly coincides with one of design points. Identity (4.1) holds for $\theta_{in}^{\text{even}}$ as well but now $\hat{j} = \hat{j}(i)$ is an odd integer.

Let i be fixed, and let the even integer $j_0 = j_0(i)$ be such that

$$(i/n_1, \theta_i) \in [X_{i,j_0-1}, X_{i,j_0+1}).$$

Define $\overline{\theta}_i^{\text{odd}}$ as the second coordinate of the midpoint of this interval, i.e.

$$(i/n_1, \overline{\theta}_i^{\text{odd}}) = (X_{i,j_0-1} + X_{i,j_0+1})/2 = X_{i,j_0}, \quad i = 1, \ldots, n_1.$$

The definition of $\overline{\theta}_i^{\text{even}}$ is quite similar, with j_0 odd.

Now we want to find the distribution functions of the random variables $n_1(\theta_{in}^{\text{odd}} - \theta_i)$ and $n_1(\theta_{in}^{\text{even}} - \theta_i)$.

Unlike the case of Lemma 3.2 these random variables are biased:

$$\liminf_{n\to\infty} \sup_{G\in\mathcal{G}(\gamma,L,h)} E_G(n_1(\theta_{in}^{\mathrm{odd}} - \theta_i)) > 0.$$

The reason is quite obvious: the distribution of $\theta_{in}^{\mathrm{odd}}$ doesn't change if θ_i, varies within the interval $((j_0 - 1)/n_1, j_0/n_1)$, j_0 even, since the corresponding points $(i/n_1, \theta_i)$ are covered by the same design interval $[X_{i,j_0-1}, X_{i,j_0+1})$ under any "shift value" $\eta_i, 0 < \eta_i < n_1^{-1}$. The same is true for $\theta_{in}^{\mathrm{even}}$: the random shift coveres only one half of each design interval. As in Lemma 3.2, we first find the distribution of the difference $n_1(\overline{\theta}_i^{\mathrm{odd}} - \theta_i)$. Unfortunately, this difference is not uniformly distributed as was the case in Lemma 3.2. Routine considerations show that the random variable $n_1(\overline{\theta}_i^{\mathrm{odd}} - \theta_i)$ is uniformly distributed in

(4.2)
$$\begin{cases} (-1, -\tau_i) \cup (1 - \tau_i, 1) & \text{if } [n_1\theta_i] \text{ is even},\\ (-\tau_i, 1 - \tau_i) & \text{if } [n_1\theta_i] \text{ is odd}, \end{cases}$$

where $\tau_i = \tau_i(\theta_i) = n_1\theta_i - [n_1\theta_i], 0 \le \tau_i < 1$. Similarly, the random variable $n_1(\overline{\theta}_i^{\mathrm{even}} - \theta_i)$ is uniformly distributed in

(4.3)
$$\begin{cases} (-1, -\tau_i) \cup (1 - \tau_i, 1) & \text{if } [n_1\theta_i] \text{ is odd},\\ (-\tau_i, 1 - \tau_i) & \text{if } [n_1\theta_i] \text{ is even}. \end{cases}$$

Hence by analogy to Lemma 3.2 one may expect that the distribution functions of $n_1(\theta_{in}^{\mathrm{odd}} - \theta_i), n_1(\theta_{in}^{\mathrm{even}} - \theta_i)$ are approximated by the distributions having piecewise constant densities. But unlike the probability density in Lemma 3.2, they have a saw-tooth shape with periodical zero gaps. The period of zero gaps is equal to 1, and the gaps for $n_1(\theta_{in}^{\mathrm{odd}} - \theta_i)$ alternate with those for $n_1(\theta_{in}^{\mathrm{even}} - \theta_i)$. Thus it is reasonable to mix the estimators $\theta_{in}^{\mathrm{odd}}$ and $\theta_{in}^{\mathrm{even}}$ randomly, i.e. to define

(4.4)
$$\theta_{in}^* = \begin{cases} \theta_{in}^{\mathrm{odd}} & \text{with probability } 1/2\\ \theta_{in}^{\mathrm{even}} & \text{with probability } 1/2 \end{cases}$$

where some independent stochastic procedure is applied to make randomization in (4.4). We use in (4.4) the same notation θ_{in}^* as that in Section 3 since only this randomized estimator is considered below.

4.1. Lemma. *The random variable $n_1(\theta_{in}^* - \theta_i)/2$ satisfies the statement of Lemma 3.2.*

Proof is the same as in Lemma 3.2: the discrete distributions

$$P_G\left(n_1(\theta_{in}^{odd} - \bar{\theta}_i^{odd})/2 = k|\mathcal{X}_i\right) \approx \pi_k$$

and

$$P_G\left(n_1(\theta_{in}^{even} - \bar{\theta}_i^{even})/2 = k|\mathcal{X}_i\right) \approx \pi_k$$

must be convolved with (4.2) and (4.3) and then mixed randomly.

Let $j^* = j^*(i)$ be such that $(i/n_1, \theta_{in}^*) = X_{i,j^*}, i = 1, \ldots, n_1$ (here j^* is randomized and is defined by θ_{in}^*). Define $Y_i^* = Y_{i,j^*}$ where Y_{i,j^*} are observations (2.2) at the design points X_{i,j^*}:

$$(4.5) \qquad Y_i^* = I(X_{i,j^*} \in G) + \zeta_i, \quad i = 1, \ldots, n_1,$$

and $\zeta_i = \xi_{i,j^*}$ are i.i.d. $(0, \sigma^2)$-Gaussian random variables. Note that ζ_i is independent of θ_{in}^*.

At the first step of the modified procedure the pairs (Y_i^*, θ_{in}^*) are calculated. It turns out that the total necessary information, additional to θ_{in}^*, is concentrated in Y_i^*: the knowledge of pairs (Y_i^*, θ_{in}^*) is sufficient to define the minimax linewise procedure.

The second step of the modified linewise procedure is determined by the LSE for piecewise-polynomial approximation. Let $\delta_n = n^{-1/(\gamma+1)}$, and $M = M_n = 1/\delta_n$ (assume without loss of generality that M is an integer). Define points $u_\ell = \ell \delta_n$ and slices $K_l = [u_{\ell-1}, u_\ell) \times R^1$, and introduce the domains $G_\ell = G \cap K_\ell, \ell = 1, \ldots, M$.

In each slice K_ℓ define a parametric family of domains

$$B_\ell(\alpha) = \left\{(x_1, x_2) \in K_\ell : 0 \leq x_2 \leq \alpha_0 + \alpha_1(x_1 - u_{\ell-1}) + \ldots + \alpha_k(x_1 - u_{\ell-1})^k\right\}$$

where $\alpha = (\alpha_0, \ldots, \alpha_k)$ is a vector of real parameters. For some values of α the curve

$$g_\ell(x_1; \alpha) = \alpha_0 + \alpha_1(x_1 - u_{\ell-1}) + \ldots + \alpha_k(x_1, u_{\ell-1})^k, u_{\ell-1} \leq x_1 < u_\ell,$$

may leave the unit square K. Let

$$A^{(K)} = \left\{\alpha : \quad 0 \leq \alpha_0 + \alpha_1(x_1 - u_{\ell-1}) + \ldots + \alpha_k(x_1 - u_{\ell-1})^k \leq 1, u_{\ell-1} \leq x_1 < u_\ell\right\}.$$

For any $\alpha \in A^{(K)}$ the domain $B_\ell(\alpha)$ is correctly defined and located within K.

Consider the MLE $\hat{\alpha}_\ell$ of α based on observations in slice K_ℓ. To avoid technical problems we consider $\alpha \in A_n^{(K)}$ where $A_n^{(K)}$ is the discrete set

$$A_n^{(K)} = \left\{ \alpha = (m_0 \delta_n^\gamma, m_1 \delta_n^{\gamma-1}, \ldots, m_k \delta_n^{\gamma-k}) : m_j \text{ integers} \right\} \cap A^{(K)}.$$

The MLE $\hat{\alpha}_\ell = (\hat{\alpha}_{0\ell}, \ldots, \hat{\alpha}_{k\ell})$ is defined as a solution of the extremal problem

$$(4.6) \qquad J_\ell(\hat{\alpha}_1) = \max_{\alpha \in A_n^{(K)}} J_\ell(\alpha)$$

where

$$J_\ell(\alpha) = \sum_i I\left(X_{i,j^*} \in B_\ell(\alpha)\right) Y_i^* - (1/2) \operatorname{card} \left\{i : X_{i,j^*} \in B_\ell(\alpha)\right\}.$$

we define the edge estimator

$$\hat{g}_\ell(x_1) = \hat{\alpha}_{0\ell} + \hat{\alpha}_{1\ell}(x_1 - u_{\ell-1}) + \ldots + \hat{\alpha}_{k\ell}(x_1 - u_{\ell-1})^k$$

for $u_{\ell-1} \le x_1 < u_\ell$, $\ell = 1, \ldots, M$. Finally, the estimator G_n^* of G is the closure of the set

$$(4.7) \qquad \{(x_1, x_2) : 0 \le x_2 \le \hat{g}_\ell(x_1), 0 \le x_1 \le 1\}$$

The following theorem states that G_n^* has the best possible rate of convergence.

4.2 Theorem. *Let θ_{in}^* be the randomized estimators (4.4) of the change-point parameters $\theta_i = g(i/n_1)$ based on observations (2.1). Then the domain estimator G_n^* defined as the closure of (4.7) has the minimax rate of convergence $\psi_n = n^{-\gamma/(\gamma+1)}$ in d-metric.*

5. Proofs

We use some results of our paper (Korostelev and Tsybakov (1992)) which we refer to in this section as KT. Assume without loss of generality that $n_1 \delta_n = n^C$ is an integer tending to infinity as $n \to \infty$; $C = (\gamma - 1)/(2(\gamma + 1))$ is positive for $\gamma > 1$. Introduce the next notation for the part of line $x_1 = i/n_1$ which belongs to the domain $B_\ell(\alpha) \Delta G_\ell$:

$$V_i = V_i(\alpha) = \{(x_1, x_2) : x_1 = i/n_1\} \cap [B_\ell(\alpha) \Delta G_\ell].$$

By definition V_i are connected intervals. Let $|V_i|$ be the length of the interval V_i, and let

$$I_\ell = \{i : u_{\ell-1} \le i/n_1 < u_\ell\}$$

be the set of indices i such that $V_i \subset K_\ell$. Note that

$$\text{card } I_\ell = n_1 \delta_n = n^C \to \infty \text{ as } n \to \infty.$$

We associate with each $\alpha \in A_n^{(K)}$ the set of integers $m = (m_0, \ldots, m_k)$ in the natural way (see the definition of $A_n^{(K)}$) and mark this correspondence by $\alpha = \alpha(m)$. Introduce the norm $\| m \|$ as maximum of $|m_j|, 0 \le j \le k$. Let the set of integers $m_\ell^{(0)}$ correspond to the Taylor approximation of edge function $g(x_1)$ around $x_1 = u_{\ell-1}$, and $\hat{m}_\ell$ correspond to the estimator $\hat{\alpha}_\ell$ in (4.6): $\hat{\alpha}_\ell = \alpha(\hat{m}_\ell)$.

The problem (4.6) is equivalent to

(5.1), $$\max_{\alpha \in A_n^{(K)}} Z_\ell(\alpha), Z_{\ell(\alpha)} = J_\ell(\alpha) - J_\ell^{(G)}$$

where

$$J_\ell^{(G)} = \sum_{i \in I_\ell} I\left(X_{i,j\bullet} \in G\right) Y_i^* - (1/2) \text{ card } \{i : X_{i,j\bullet} \in G\}$$

is independent of α.

5.1. Lemma. *For each* $\ell, \ell = 1, \ldots, M$, *the random function* $Z_\ell(\alpha)$ *can be presented in the form*

$$Z_\ell(\alpha) = \Delta_\ell(\alpha) - (1/2) \sum_{i \in I_\ell} P_G\left(X_{i,j\bullet} \in V_i(\alpha)\right)$$

where the random variables $\Delta_\ell(\alpha)$ are independent for different ℓ, $E_G[\Delta_\ell(\alpha)] = 0$, and for all z small enough

(5.2) $$E_G[\exp(z\Delta_\ell(\alpha))] \leq \exp\left\{\lambda_0 z^2 \sum_{i \in I_\ell} P_G(X_{i,j\bullet} \in V_i(\alpha))\right\}$$

with a positive λ_0 which does not depend on n, ℓ, α and G.

Proof. Note that (4.5) entails the representation

$$Z_\ell(\alpha) = \sum_i I(X_{i,j\bullet} \in B_\ell(\alpha) \setminus G_\ell)\zeta_i - \sum_i I(X_{i,j\bullet} \in G_\ell \setminus B_\ell(\alpha))\zeta_i - (1/2)\Lambda_\ell =$$

$$= \Delta_\ell(\alpha) - (1/2)\Lambda_\ell$$

where

$$\Delta_\ell(\alpha) = \sum_i I(X_{i,,j\bullet} \in B_\ell(\alpha) \setminus G_\ell)\zeta_i - \sum_i I(X_{i,j\bullet} \in G_\ell \setminus B_\ell(\alpha))\zeta_i$$

and $\Lambda_\ell = \text{card }\{i : X_{i,j\bullet} \in B_\ell(\alpha)\Delta G_\ell\}$ with the mean value

$$E_G(\Lambda_\ell) = \sum_{i \in I_\ell} P_G(X_{i,j\bullet} \in V_i(\alpha)).$$

The rest of proof is similar to that in KT, Lemma 4.2.2. ∎

5.2. Lemma. For n large enough and $\alpha \in A_n^{(K)}$

$$(1/2)d(B_\ell(\alpha), G_\ell) \leq (1/n_1) \sum_{i \in I_\ell} |V_i(\alpha)| \leq 2d(B_\ell(\alpha), G_\ell)$$

where $n_1 - n^{1/2}$.

Proof. Note that $(1/n_1) \sum_{i \in I_\ell} |V_i(\alpha)|$ is the Riemann sum for

$$\int_{u_{\ell-1}}^{u_\ell} |\alpha_0 + \alpha_1(x_1 - u_{\ell-1}) + \ldots + \alpha_k(x_1 - u_{\ell-1})^k - g(x_1)|dx_1 = d(B_\ell(\alpha), G_\ell).$$

The latter quantity is approximated by the Riemann sum with accuracy

$$O(d(B_\ell(\alpha), G_\ell)/\text{ card } I_\ell) \leq (1/2)d(B_\ell(\alpha), G_\ell)$$

for all sufficiently large n. ∎

5.3. Lemma. *If* $\| m - m_\ell^{(0)} \| \leq n^C/(\gamma + 1)$ *and* $\alpha = \alpha(m)$, *then for all* ℓ, $\ell = 1, \ldots, M$, *and* n *large enough the following inequalities hold:*

(5.3)
$$\max_{i \in I_\ell} |V_i(\alpha)| \leq 1/n_1;$$

(5.4)
$$n_1(C_1|V_i(\alpha)| - \exp(-\bar{\lambda}n_1)) \leq P_G(X_{i,j\cdot} \in V_i(\alpha)) \leq n_1(C_2|V_i(\alpha)| + \exp(-\bar{\lambda}n_1))$$

(5.5)
$$(1/2)C_1 d(B_\ell(\alpha), G_\ell) - \exp(-\bar{\lambda}n_1) \leq \sum_{i \in I_\ell} P_G(X_{i,j\cdot} \in V_i(\alpha))$$

$$\leq 2C_2 d(B_\ell(\alpha), G_\ell) + \exp(-\bar{\lambda}n_1),$$

with some positive constants C_1, C_2, *and* $\bar{\lambda}$.

Proof. The boundaries of $B_\ell(\alpha)$ and G_ℓ are given respectively by the edge function $g_\ell(x_1; \alpha) = \delta_n^\gamma(m_0 + m_1 y + \ldots + m_k y^k)$ where $y = (x_1 - u_{\ell-1})\delta_n$, $0 \leq y \leq 1$, and by $g(x_1)$. Comparing this whith the Taylor expansion for $g(x_1)$ at $x_1 = u_{\ell-1}$, we get

$$\max_{i \in I_\ell} |V_i(\alpha)| \leq \delta_n^\gamma L/k! + \delta_n^\gamma(k+1) \| m - m_\ell^{(0)} \| \leq \delta_n^\gamma L/k! + \delta_n^\gamma n^C(k+1)/(\gamma+1) \leq$$

$$\leq \delta_n^\gamma n^C = n^{-1/2} = 1/n_1,$$

for n large, which proves (5.3).

Note that

$$P_G(X_{i,j\cdot} \in V_i(\alpha)) = P_G[n_1(\theta_{in}^* - \theta_i)/2 \in (n_1/2)(V_i(\alpha) - \theta_i)]$$

where $(V_i(\alpha) - \theta_i)$ is the shift of the interval $V_i(\alpha)$. By definition, one of the ends of $(V_i(\alpha) - \theta_i)$ is the origin. Hence, Lemma 4.1 applies which states that the distribution function of $(n_1/2)(\theta_{in}^* - \theta_i)$ is approximated by that having density $\pi(x)$. Taking into account (5.3) one gets (5.4). The inequalities (5.5) are the direct consequence of Lemma 5.2 and (5.4). ∎

5.4 Lemma. *If* $\| m - m_l^{(0)} \| > n^C/(\gamma + 1)$ *and* $\alpha = \alpha(m) \in A_n^{(K)}$, *then*

(5.6)
$$\sum_{i \in I_\ell} P_G \left(X_{i,j\bullet} \in V_i(\alpha(m)) \right) \geq C_3 n^C$$

for n large enough with some constant $C_3 > 0$.

Proof. Assume for definiteness that $g(i/n_1) \leq g_\ell(i/n_1; \alpha(m))$. Then

$$P_G(X_{i,j\bullet} \in V_i(\alpha)) = P_G(g(i/n_1) \leq \theta_{in}^* \leq g_\ell(i/n_1; \alpha(m)))$$

Using the definition of these functions, we have

$$g_\ell(x_1; \alpha(m)) - g(x_1) = \left[\sum_{j=0}^{k} \left[\alpha_j(m) - \frac{g^{(j)}(u_{l-1})}{j!} \right] (x_1 - u_{l-1})^j \right] +$$

$$+ \left[\sum_{j=0}^{k} \frac{g^{(j)}(u_{l-1})}{j!} (x_1 - u_{l-1})^j - g(x_1) \right], \quad u_{l-1} \leq x_1 < u_l.$$

Since the last summand presents the Taylor approximation, we have

$$g_\ell(x_1; \alpha(m)) - g(x_1) \geq \delta_n^\gamma [\sum_{j=0}^{k} (m_j - m^{(0,j)}) y^j - L/k!] \geq$$

$$\geq \delta_n^\gamma \| m - m_\ell^{(0)} \| \left[\sum_{j=0}^{k} (m_j - m_\ell^{(0,j)}) \| m - m_\ell^{(0)} \|^{-1} y^j - L/k! \| m - m_\ell^{(0)} \|^{-1} \right]$$

where $y = (x_1 - u_{l-1})/\delta_n$. Since $\| m - m_\ell^{(0)} \| > n^C/(\gamma + 1)$ the last term is negligible as n tends to infinity. There exists an interval U of length $|U| > 0$ and a positive C_4 such that

$$\sum_{j=0}^{k} (m_j - m^{(0,j)}) \| m - m_\ell^{(0)} \|^{-1} y^j \geq C_4 \text{ for } y \in U.$$

Indeed, the set of polynomials on the left-hand side is compact and separated from zero. It means that among $\{V_i(\alpha(m)), i \in I_l\}$ in (5.6) there exist at least $|U| n_1 \delta_n = |U| n^C$ intervals with lenghts exceeding

$$(C_4/2) \delta_n^\gamma n^C/(\gamma + 1) \geq C_4/2 \, \delta_n^{\gamma+1} n_1/(\gamma + 1) = C_4/2 \, n_1^{-1}/(\gamma + 1).$$

For n large each of the corresponding probabilities $P_G(X_{i,j^*} \in V_i(\alpha(m)))$ is bounded from below by a positive constant in accordance with (5.4). This proves (5.6). ∎

Proof of Theorem 4.2. It was shown in KT, theorem 4.1.1, that Lemmas 5.1, 5.2, and inequality (5.5) entail inequality for any $R > 0$:

(5.7)
$$P_G \left[\max_{R \leq \|m - m_\ell^{(0)}\| \leq n^C/(\gamma+1)} [Z_\ell(\alpha(m)) - Z_\ell(\alpha(m_\ell^{(0)}))] > 0 \right] \leq C_5 \exp(-C_6 R).$$

with some positive constants C_5, C_6.

For each m such that $\| m - m_\ell^{(0)} \| > n^C/(\gamma + 1)$ Lemma 5.4 guarantees that the corresponding probability $P_G(\hat{\alpha}_\ell = \alpha(m))$ decreases exponentially as $n \to \infty$. In fact, for n large enough we have

$$P_G(\hat{\alpha}_\ell = \alpha(m)) \leq P_G(Z_\ell(\alpha(m)) - Z_\ell(\alpha(m_\ell^{(0)}))) > 0) \leq$$

$$\leq P_G \left[\Delta_\ell(\alpha(m)) - \Delta_\ell(\alpha(m_\ell^{(0)})) > (1/4) \sum_{i \in I_\ell} P_G(X_{i,j^*} \in V_i(\alpha)) \right] \leq$$

$$\leq C_7 \exp \left[\lambda_0 z^2 \sum_{i \in I_\ell} P_G(X_{i,j^*} \in V_i(\alpha)) - (z/4) \sum_{i \in I_\ell} P_G(X_{i,j^*} \in V_i(\alpha)) \right]$$

where $C_7 > 0$ is a constant and λ_0 is the constant from (5.2). If we take $z = (1/8\lambda_0)$ and apply Lemma 5.4, then

$$P_G(\hat{\alpha}_\ell = \alpha(m)) \leq C_7 \exp \left\{ -C_8 \sum_{i \in I_\ell} P_G(X_{i,j^*} \in V_i(\alpha)) \right\} \leq C_7 \exp(-C_8 n^C/(\gamma+1))$$

where $C_8 = (64\lambda_0)^{-1}$. Hence

(5.8)
$$P_G \left(\| \hat{m}_\ell - m_\ell^{(0)} \| > n^C/(\gamma + 1) \right) \leq$$

$$\leq C_7 \operatorname{card} (A_n^{(K)}) \exp(-C_8 n^C/(\gamma + 1)) \leq \exp(-C_9 n^C)$$

for $C_9 = C_8/2$ and n large enough. Note that $\| \hat{m}_\ell - m_\ell^{(0)} \|$ are independent random variables, $\ell = 1, \ldots, M$, and in accordance with (5.7) and (5.8) their distribution have exponentially decreasing tails.

It is shown in KT, Lemma 4.2.1, that

$$d(B_\ell(\alpha(m)), G_\ell) \leq \delta_n^{\gamma+1}(\lambda_3 \| m - m_\ell^{(0)} \| + \lambda_4)$$

131

with some positive constants λ_3, λ_4. Hence

$$(5.9) \qquad P_G(\delta_n^{-\lambda} d(G, G_n^*) \geq q) \leq P_G\left[\delta_n \sum_{\ell=1}^{M} (\lambda_3 \parallel \hat{m} - m_\ell^{(0)} \parallel + \lambda_4) \geq q\right] \leq$$

$$\leq P_G\left[M^{-1} \sum_{\ell=1}^{M} \parallel \hat{m} - m_\ell^{(0)} \parallel \geq (q - \lambda_4)/\lambda_3\right].$$

The last probability decreases exponentially in q as $q \to \infty$. This proves the theorem. ∎

REFERENCES

Blake,A. and Zisserman,A. : (1987) *Visual Reconstruction*. MIT Press, Cambridge, MA.

Bretagnolle,J. and Huber,C. (1979) Estimation des densités: risque minimax, *Z. für Wahrscheinlichkeitstheorie und verw. Geb* 47, 119-137.

Geman,S. and Geman,D. (1984) Stochastic relaxation, Gibbs distribution, and Bayesian restoration of images. *IEEE Trans. on PAMI*, 6, 721-741.

Girard,D. (1990) From template matching to optimal approximation by piecewise smooth curves. In: *Curves and Surfaces in Computer Vision and Graphics*, Proc. Conf. Santa Clara, California, 13-15 february, 1990, 174-182.

Haralick,R.M.(1980) Edge and region analysis for digital image data.*Comput. Graphics and Image Processing* 12, 60-73.

Härdle,W.(1990) *Applied Nonparametric Regression*, Cambridge Univ. Pres.

Huang,J.S. and Tseng, D.H. (1988) Statistical theory of edge detection. *Comput. Graphics and Image Processing* 43, 337-346.

Ibragimov,I.A. and Khasminskii,R.Z. (1979) *Statistical Estimation: Asymptotic Theory*, Springer-Verlag, N.Y.

Korostelev,A.P. (1991) Minimax reconstruction of two-dimensional images, *Theory of Probability and its Applications* 36, 153-159.

Korostelev,A.P. and Tsybakov,A.B. (1992) *Asymptotically minmax image reconstruction problems*. In: "Topics in Nonparametric Estimation", Khasminskii R.Z., ed., AMS, Providence, RI.

Marr,D.(1982) *Vision*, W.H. Freeman and Co., San Francisco.

Mumford,D. and Shah,J.(1989) Optimal approximation by piecewise smooth functions and associated variational problems. *Comm. Pure Appl. Math.* 42, 577-685.

Nagao,M. and Matsujama,T. (1979) Edge preserving smoothing. *Graphics and Image Processing* 9, 394-407.

Pratt,W.K.(1978) *Digital Image Processing*, J. Wiley, N.Y.

Ripley,B.D. (1988) *Statistical Inference for Spatial Processes*. Cambridge Univ. Press.

Rosenfeld,A. and Kak,A.C. (1982) *Digital Picture Processing*. Academic Press, London.

Shiau,J.J., Wahba,G. and Johnson,D.R. (1986) Partial spline models for the inclusion of tropopause and frontal boudary information in otherwise smooth two- and three-dimensional objective analysis. *J. Atmospheric and Oceanic Technology* 3, 714-725.

Stone,C.J. (1980) Optimal convergence for nonparametric estimators, *Annals of Statistics* 8, 1348-1360.

Stone,C.J. (1982) Optimal global rates of convergence for nonparametric estimators, *Annals of Statistics* 10, 1040-1053.

Titterington,D.M. (1985) Common structure of smoothing techniques in statistics. *Internat. Statist. Review* 53, 141-170.

Torre,V. and Poggio,T. (1986) On edge detection. *IEEE Trans. on PAMI* 8, 147-163.

Tsybakov,A.B. (1989) Optimal estimation accuracy of nonsmooth images, *Problems of Information Transmission* 25, 180-191.

Bandwidth Selection for Kernel Regression: a Survey

P. Vieu

Laboratoire de Statistique et Probabilités, Université Paul Sabatier, URA CNRS D745, 31062 Toulouse Cedex, France.

Key Words: Asymptotic optimality; Bandwidth choice; Bibliography; Bootstrap; Cross-validation; Kernel regression; Plug-in.

Abstract

This paper is concerned with nonparametric estimation of a regression function. The behaviour of kernel estimates depends on a smoothing parameter (i.e. the bandwidth). Bandwidth choice turns out to be of particular importance as well for practical use as to insure good asymptotic properties of the estimate. Various techniques have been proposed in the past ten last years to select optimal values of this parameter. This paper presents a survey on theoretical results concerned with bandwidth selection.

1. Introduction

Starting with the definition of kernel regression estimates (Watson (1964) and Nadaraya (1964)), the importance of nonparametric techniques in regression estimation has been considerably encreased during the past thirthy years. Until the middle of the eighties, theoretical investigations in nonparametric regression were concentrated on showing consistency properties for several different classes of estimates (see Collomb 1981 and 1985 for surveys). This was an important field of statistical researches, and more than two hundreds papers have been collected in Collomb's surveys. Nonparametric regression estimates involve a smoothing parameter, and main interest of these consistency results is to give light on the prominent effect of this parameter. Then, researches in nonparametric statistics focused on the natural question: How one has to choose this smoothing factor in practical situations?

Previously, one was choosing the smoothing parameter just by trying several values and by selecting "by eyes" the value leading to the "best" estimated curve. It was of course necessary to have some more precise way to do that, at least to make the use of nonparametric methods accessible to everyone. Data-driven bandwidth selection rules became then one of the main purpose of researches in nonparametric statistics. Starting with the paper by Härdle and Marron (1985a) showing optimality of Cross-Validation selection rule (see Section 4.1), several works have been published on this subject over this past decade. We propose in this paper a survey on the theoretical results concerning smoothing parameter selection, and we decided to restrict our attention on the conceptually simple kernel estimates. We concentrate only on theoretical aspects of the problem. Concerning applications to several different data sets (real or simulated), we encourage the reader in looking at the monograph by Härdle (1990).

This paper is organized as follows. Section 2 recalls basic definitions of kernel estimates. In Section 3 we define several different kinds of optimality properties that a data-driven bandwidth will be expected to satisfy. Main section of this paper is Section 4 in which usual selection methods are presented together with their main optimality properties. One attempt of this survey was to emphasize the current limitations of the results to be the starting point of new research projects. This will be discussed in the concluding Section 5.

2. The smoothing parameter selection problem

2.1. Several versions of kernel estimates

Assume that we have a sample (X_i, Y_i), $i=1,\dots,n$ of realizations of a pair (X,Y) where $X \in \mathbb{R}^p$ and $Y \in \mathbb{R}$. Consider the problem of the estimation of the following regression function

$$Y_i = r(X_i) + \varepsilon_i, \text{ for any } i=1,\dots,n,$$

where ε_i are i.i.d. random variables with zero mean. Three different situations are considered according to the randomness of the explanatory variable X. Each situation leads to a different type of kernel estimate.

Random design with unknown marginal density. Assume that X is a random variable with unknown density f. Kernel estimates were introduced by

136

Watson (1964) and Nadaraya (1964). They are defined from a kernel function K on $\mathbb{R}^p$ and from a bandwidth h (function of the sample size n) by

$$r_{h,1}(x) = [\sum_{i=1}^{n} Y_i K((x-X_i)/h)] / [\sum_{i=1}^{n} K((x-X_i)/h)].$$

Random design with known marginal density. In some situations the marginal density f of X is known and therefore a simpler form of kernel estimate has been introduced by Johnston (1982):

$$r_{h,2}(x) = \frac{1}{nh^p} [\sum_{i=1}^{n} Y_i K((x-X_i)/h)] / f(x).$$

Of course such a case is very much restrictive than when f is unknown. In a technical point of view this estimate is easier to deal with because it does not involve random denominator. This will be the main reason why theoretical investigations are deeper in this situation than in the previous case.

Fixed design points. Even simpler is the situation when the X_i are not random but are fixed design points. This may occure in several practical situations, for instance when the explanatory variable X is time. Theory concerning this case is always carried out for real variable X, (i.e. for p=1). Also, it is always assumed that the X_i are increasing with i and that they are valued in the interval [0,1]. In the remaining of this paper such restrictions will be implicitly assumed when dealing with fixed design regression. Gasser and Müller (1984) give motivations for using the following convolution type of kernel estimate:

$$r_{h,3}(x) = \frac{1}{nh} \sum_{i=1}^{n} \int_{s_{i-1}}^{s_i} Y_i K((x-u)/h)du,$$

where $s_0=0$, $s_1=n$ and $s_i=(X_i+X_{i+1})/2$. When the design points are equally spaced (i.e. $X_i=i/n$), $r_{h,3}$ is just the estimate $r_{h,2}$ in which f is taken to be uniform on the interval [0,1]. In fact, both are also equal to the estimate introduced by Priestley and Chao (1972).

Comparison of these various kinds of kernel estimate is carried out in the recent papers by Chu and Marron (1991b) and Jones and Davies (1991). In order to avoid the troubles caused by the random denominator in $r_{h,1}$, Mack and Müller (1989a, 1989b) define a convolution type estimator for the random design situation. Similarly, using earlier ideas of Lejeune (1985), Fan (1992a, 1992b, 1992c) defines a new kernel type estimate for the random design case.

Both Fan and Mack-Müller's aproaches are very promising, but bandwidth selection has still not been investigated for these estimates until now.

2.2. The bandwidth choice problem

Each kind of kernel estimate involves the bandwidth h that controls the degree of smoothing of the estimated curve. Let r_h be one among the estimates $r_{h,1}$, $r_{h,2}$ or $r_{h,3}$. For instance, the role of h can be seen through Integrated Square Errors (ISE)

$$ISE(h) = \int (r_h(x) - r(x))^2 w(x)f(x)dx.$$

First results about asymptotic evaluation of ISE was given by Parzen (1962) for density estimation, and Collomb (1976) extended them to the regression setting. We have the following result

$$ISE(h) = \frac{C_1}{nh^p} + C_2 h^{2k} + o_p(\frac{1}{nh^p} + h^{2k}). \qquad (2.1)$$

In these expressions w is a weight function to be specified later on, k is the number of continuous derivatives one has to assume for the functions f and r, and C_1 and C_2 are two finite real numbers which are independent of h but depend only on K and on the derivatives of f and r. We do not give here expressions of these constants since it would need long and tedious notations, because C_1 and C_2 differ according to the form of the kernel estimate in consideration. The reader will find details on them in Vieu (1991a) for both first estimates and in Gasser et al (1991) for the third one. Let us mention the additional work by Hall (1984b) in which asymptotic normal distribution of ISE is investigated.

First important consequence of (2.1) is that kernel estimates may reach the Stone's optimal global rate of convergence. To see that, compute the bandwidth that minimizes the leading terms in the expression (2.1), i.e:

$$h_1 = (\frac{pC_1}{2kC_2})^{1/(2k+p)} n^{-1/(2k+p)}, \qquad (2.2)$$

and note that $ISE(h_1)$ is of order $n^{-2k/(2k+p)}$. This rate of convergence is known to be optimal over the class of k times continuously differentiable regression functions. These optimal rates were given by Stone (1982) (see also Konakov and Piterbarg (1984) and Hall (1989) for complementary results). Unfortunately the bandwidth h_1 is unusable in practice because it depends on the quantities C_1 and C_2 that involve the unknown functions r and f and their derivatives.

At this stage, one had to look for data-driven bandwidth $\hat{h}$ that could be used in practice. Naturally, a data-driven bandwidth is expected to estimate correctly the **theoretical asymptotic optimal bandwidth** h_1. Even better would be to estimate correctly the **theoretical optimal bandwidth** h_2:

$$h_2 = \underset{h \in H_n}{\arg \min} \ ISE(h),$$

where H_n is a set of possible bandwidths to be precised below. Note that estimating h_2 may be done under more general nonparametric model since it will not need any assumption on the number k of derivatives of r, while estimating h_1 did since it will depend crucially on the result (2.1).

Before closing this section let us note that our paper is presented in terms of ISE, but in fact all remains true when ISE is changed in the discrete quadratic measure of errors:

$$ASE(h) = \frac{1}{n} \sum_{i=1}^{n} (r(X_i) - r_h(X_i))^2 w(X_i),$$

or in the Mean Integrated Square Errors

$$MISE(h) = E(ISE(h)).$$

These three quadratic measures are asymptotically equivalent for a very large class of estimates including all the estimates investigated herein (Marron and Härdle (1986)). In a recent paper Härdle et al (1992) note that it is a little bit different when one is dealing with

$$MASE(h) = E(ASE(h)).$$

3. Optimality properties of bandwidth selection rules

The first natural property a selected bandwidth $\hat{h}$ has to satisfy is:

$$\frac{\hat{h}}{h_1} \rightarrow 1. \tag{3.1}$$

Such an approach has the drawback to depend on h_1 and so to need assumption about existency of derivatives for r (see (2.2)). To avoid this difficulty one would rather prefer to show property of the form

$$\frac{\hat{h}}{h_2} \rightarrow 1. \tag{3.2}$$

In fact, it will be seen in Section 4 that each approach will lead to different kind of bandwidth selection rules. A more interesting property is the following notion of asymptotic optimality:

$$\frac{ISE(\hat{h})}{ISE(h_2)} \rightarrow 1. \tag{3.3}$$

In (3.1)-(3.3) the kind of convergence is not precised, and in the remaining, of our paper such properties could be given alternatively almost surely, in probability or with complete convergence.

Let us now introduce basic assumptions that are used by most of authors. These conditions are those introduced by Härdle and Marron (1985) when they proved the first result of type (3.3).

$$\text{The pairs } (X_i, Y_i) \text{ are independent;} \tag{A.1}$$

$$H_n = [C^{-1} n^{\delta - 1/p}; C n^{-\delta}] \text{ for some } \delta, \ 0 < \delta < 1/p; \tag{A.2}$$

$$K, r \text{ and } f \text{ are Hölder continuous;} \tag{A.3}$$

$$\forall q > 0, \ \exists A_q > 0 \ E(|Y|^q | X = x) < A_q \text{ for all } x; \tag{A.4}$$

$$w \text{ has a compact support } S; \tag{A.5}$$

$$f \text{ is bounded from below on the interior of } S. \tag{A.6}$$

4. Three different approaches of bandwidth selection

4.1. Cross-Validation and related methods

The most part of theoretical results on bandwidth selection is concerned with Cross-Validation technique. Main idea is derived from model selection procedure. It consists in defining a score function which has asymptotically the same shape as the function ISE(h) up to a constant term (i.e. a term independent of h). Both functions are then expected to have the same asymptotic minimisers and so optimality properties presented in Section 3 are expected to hold. Let r_h be one among the three versions of kernel estimates defined in Section 2. The **Cross-Validation score function** is defined by

$$CV(h) = \frac{1}{n} \sum_{i=1}^{n} (Y_i - r_h^{-i}(X_i))^2 w(X_i),$$

where r_h^{-i} is the leave-out kernel estimate based on the sample in which the ith observation (X_i, Y_i) has been deleted. The bandwidth selection procedure consists in choosing

$$\hat{h} = \arg\min_{h \in H_n} CV(h).$$

Before giving theoretical results, let us just say something about the existency of such a minimum. In fact, for finite sample size it is quite often the case in practice when the score function presents multiple local minima. This is indeed the less attractive feature of this method. In regression, this has not received any theoretical explanations. We guess however that investigations and conclusions presented by Hall and Marron (1991) in density estimation will remain basically the same in regression.

Case of random design with unknown marginal density. Here r_h is the Watson-Nadaraya estimate $r_{h,1}$. In Hall (1984a) the property (3.2) is shown almost surely for the particular kernel $K = 1_{[-1/2, 1/2]}$. From Hall's result, Collomb et al. (1986) derived the almost sure pointwise consistency of the Watson-Nadaraya estimate using this cross-validated bandwidth. Pointwise quadratic consistency is shown with other techniques by Wong (1982). Note that such consistency results cannot be obtained directly from (2.1) since the cross-validated bandwidth is now a random variable.

First really interesting result was given by Härdle and Marron (1985a) when they proved property (3.3) with almost sure convergence under the conditions (A.1)-(A.6). These authors gave their result with almost sure convergence, but in fact they proved it for complete convergence. Then Burman and Chen (1989) extended this result when assuming only that r is bounded (instead of (A.3)). A second nice aspect of Burman-Chen's paper is to deal with a larger class of estimates (including kernels). In Härdle and Vieu (1991), property (3.3) is given with complete convergence when the independence assumption (A.1) is relaxed. This last paper made the restrictions that r has two continuous derivatives and that p=1 (see however Vieu (1991b) for extension to the multivariate case).

The above discussed criterion is usually called ordinary Cross-Validation. It can be modified to improve performances of the selected bandwidth. For instance, a local version of the criterion CV is proposed in Vieu (1991c) and a result like (3.3) is proved when a pointwise quadratic measure of error is used instead of ISE. A robust version of Cross-Validation is introduced by Hall and

Jones (1990). Another modification consists in deleting more than one point in the sample to construct the leave out estimate (Härdle and Vieu (1991)). It is also possible to take into account the dependence without deleting too much points in the leave-out estimate, just by introducing a timing smoother (Vieu (1991b)). These two last modifications are of particular interest in case of dependent errors (but not only), and (3.3) was shown to be true for these modified Cross-Validation rules without assuming (A.1). Timing smoothing technique is of particular interest for high dependence and small sample size (see the discussion for density estimation in Hart and Vieu (1990)).

Case of random design with known marginal density. Here r_h is the estimate $r_{h,2}$. Of course the problem is in this situation much more simpler because we do not have to worry with random denominator. Therefore all the properties presented before for $r_{h,1}$ remain true for $r_{h,2}$. Additional properties can be stated. Härdle and Kelly (1987) show (3.3) when the minimization is taken over the whole positive real numbers instead of (A.2).

Several other criteria have been defined. All are closed to CV and are also derived from model selection procedures. Among them are Akaike's criterion (Akaike 1974), FPE criterion (Akaike 1970), Shibata's selector (Shibata 1981) and Mallow's selector (Rice 1984). We do not present them in details here because of the following result. Härdle et al (1988) had shown that as well CV as any among these criterions can be written as:

$$[\frac{1}{n} \sum_{i=1}^{n} (Y_i - r_h(X_i))^2 w(X_i)](1+\Phi(h)).$$

The function Φ differs according to which criterion is considered, but all these functions have the same first order Taylor expansion. Therefore, optimality of these selectors follows directly from optimality of CV. Moreover, Härdle and Marron (1985b) give several arguments to compare these criterions. They conclude that CV has the advantage to satisfy (3.3) independently of the choice of the weight function w, and that is the reason why in practice other criteria are not so often used as CV does.

The following result was also given by Härdle et al (1988). If conditions (A.1)-(A.6) hold, if f and r are assumed to have two continuous derivatives and if p=1, they show:

$$\frac{\hat{h} - h_2}{h_2} = O_p(n^{-1/10}). \qquad (4.1)$$

A more precise result about the limiting distribution of the difference $\hat{h} - h_2$ is also given by these authors. Result (4.1) shows the slow rate of convergence of cross-validated bandwidth. Further insight about that is carried out by Chiu (1989, 1990) and Härdle et al (1992).

Case of fixed design. In the fixed design setting more theoretical investigations are available because of simplicity of the model and of the estimates. This is particularly true about results for dependent data. The negative correlation between the ordinary cross-validated bandwidth $\hat{h}$ and h_2 is studied by Chiu and Marron (1990). Following previous ideas from Marron (1987), Chu and Marron (1991a) discuss the so-called partionned Cross-Validation procedure. Dividing data in subgroups, they compute the values of CV over each subgroup and they obtain a selection criterion by averaging all these CV values. They show how this partitioning technique may be interesting for situation when the data are not independent, and they compare it with the method of Härdle and Vieu (1991). Altman (1990) proposes another modification of the Cross-Validation score function which consists in adding a penalty term to take into account the dependence, and shows result (3.3) for this penalysed Cross-Validation selection rule.

4.2. Plug-in method

While selectors presented in Section 4.2 were essentially based on the estimation of the theoretical optimal bandwith h_2, there is another way to construct selected bandwidth by estimating from (2.2) the theoretical asymptotically optimal bandwidth h_1. Such kind of approach is usually known as **Plug-in selection rule.** In the setting of regression estimation there is not a lot of works in this direction. This may appear quite strange since in other functional estimation fields it is not the case. This comes mainly from technical difficulties. Estimating h_1 needs to estimate the constants C_1 and C_2 (see (2.1)) and therefore to estimate the derivatives of f and r. In other functional estimation problems this can be done without too much troubles. But in regression, even if the derivatives of Watson-Nadaraya estimates have nice optimality properties (see Györfi et al 1989 for instance), they are very much difficult to use in practice because they are defined as a ratio. For the same reason, the papers we have collected are only concerned with the case of fixed design regression.

Case of fixed design. It is always assumed that the design points X_i are equally spaced and so, under this assumption, both estimates $r_{h,2}$ and $r_{h,3}$ are equal. Introducing estimates $\hat{C}_1$ and $\hat{C}_2$ of C_1 and C_2 needs to estimate the second derivative of r. This can be done using higher order kernel techniques as presented in Gasser and Müller (1984), and by taking a bandwidth h_d for derivative estimation which is of the form

$$h_d = h_0 n^{-1/10},$$

where h_0 is an arbitrary bandwidth. Then a selected bandwidth is obtained by

$$\hat{h} = (\frac{p\hat{C}_1}{2k\hat{C}_2})^{1/(2k+p)} n^{-1/(2k+p)}.$$

Previous papers on this subject are from Müller (1985) and Staniswallis (1989). Goldstein and Messer (1990) show consistency of the kernel estimate using such a bandwidth. Gasser et al (1991) (see also Gasser and Herrmann (1991)), obtain the following rate of convergence:

$$\frac{\hat{h} - h_2}{h_2} = O_p(n^{-1/10}). \tag{4.2}$$

This result shows that Plug-in rule has the same power as CV procedure with the same drawback to have a very poor rate of convergence. Note however that they prove that the variability is smaller in (4.2) than what it is in (4.1).

There is in Gasser et al (1991) a nice idea to improve this slow rate of convergence. It consists in iterating the procedure. In a first step Plug-in method gives a selected bandwidth. Then they construct another bandwidth h_d by using this selected bandwidth instead of the arbitrary bandwidth h_0. After a short number of iterations they obtain the following rate of convergence for the iterated plug-in bandwidth $\hat{h}$:

$$\frac{\hat{h} - h_2}{h_2} = O(n^{-1/5}) + O_p(n^{-1/2}). \tag{4.3}$$

Obvious corollaries of (4.2) and (4.3) are optimality properties (3.1) and (3.2). However, asymptotic optimality results like (3.3) has not been proved for Plug-in bandwidth. Until now we just have the following result which is derived from Theorem 3 in Gasser et al. (1991):

$$\frac{ISE(\hat{h})}{ISE(h_2)} = O_p(1).$$

Let us mention another work by Herrmann et al (1991) which is devoted to extensions of this Plug-in method to the case of dependent data. In Mack and Wulwick (1991) similar ideas are given and comparison with Cross-Validation procedure is carried out through a real example. Similar approach has been investigated in a paper by Eubank and Schucany (1990). Starting from formulas similar to (2.1) and (2.2) but when ISE is changed in a pointwise quadratic measure of error, they observe that optimal bandwidth given in (2.2) leads to error for which the variance (i.e. the first term in (2.1)) is exactly 2k times the squared bias (i.e. the second term in (2.1)). Therefore, they propose to choose as bandwidth the solution of the equation

$$\text{Variance}(h) = 2k\text{Bias}(h)^2.$$

This is indeed quite close to the above mentionned Plug-in approach since computing variance and squared bias needs to estimate constants C_1 and C_2. Eubank and Schucany show local versions of consistency result (3.1) without having to use any pilote estimate.

4.3. Bootstrap for bandwidth selection

Since a few years ago Bootstrap techniques have received again a lot of attention. Their possible use for nonparametric inference is certainly one reason for that. From the previous paper by Härdle and Bowman (1987), **Bootstraped bandwidths** have been widely studied in the regression setting. Bootstrap attempts to estimate the distribution of an error of estimation (for instance ISE). Once ISE(h) is estimated, the Bootstraped bandwidth is defined to be the minimizer of the estimated ISE. Of course other measures of errors than ISE can be used and other applications of Bootstrap are available (for instance construction of confidence bands) but it is not our goal to discuss them here. (see Härdle (1990) and the references therein). As it was the case for Plug-in rules, Bootstrap has been investigated mainly for fixed design regression.

Case of fixed design. Bootstrap is working as follows. Compute the estimate r_h and derive estimates of the residuals ε_i. Then, center the estimated residuals to get $\hat{\varepsilon}_i$, i=1,...,n, and randomly generate new residuals from the set of $\{\hat{\varepsilon}_i,$ i=1,...,n$\}$. This procedure allows finally to create bootstraped data. Then we can compute a new nonparametric estimate $r_h{}^*$ of the same type as r_h just by using these bootstraped data. So one has at hand an estimate ISE*(h) of ISE(h). Härdle and Bowman (1987) propose to use as selected bandwidth:

$$\hat{h} = \underset{h \in H_n}{\arg\min} \ \text{ISE*}(h),$$

and show that this selected bandwidth satisfies the optimality property (3.3), under the restriction that r has two continuous derivatives. In fact, because estimated residuals are not unbiased an additional term (involving estimation of the second derivative of r) was previously included by Härdle and Bowman in the quantity ISE*. We do not discuss this point in details here since Härdle and Mammen (1990) had shown that this additional term can be suppressed just by overestimating the second derivative of r. An important feature of this procedure iss that all the computations are in fact made pointwisely. That means that dealing with Integrated errors is not a great advantage (while for Cross-Validation techniques discussed in Section 4.1 it did). So, as pointed out by Härdle and Bowman, their optimality result is in fact valid when ISE is changed in a pointwise measure of errors between r_h and r, and so this bandwidth selection rule is in fact location adaptive.

Similar results were given by Hall (1990) for general Lp distances but with the restriction that K is an uniform kernel function. Note also that Härdle and Marron (1990) discuss how extend Bootstrap procedure to deal with situation when ε_i depends on the location of X.

Case of random design with unknown marginal density. In the situation of random design, the Bootstrap procedure described above may cause some troubles since as mentioned by Härdle and Mammen (1990) it cannot take into account the stochastic nature of the model. Cao-Abad (1991) proposes a smooth Bootstrap procedure that consists in generating new residuals according to a smooth estimate of the bivariate distribution of (X,Y). Such an approach is clearly very appealing, but until now application to bandwidth selection has still not been investigated.

5. Concluding section

It would be interesting to close this study by replying the natural question: **Which among these selection rules has to be used?** Unfortunately, at this stage of researches in this area it is not possible to give definitive answer to this question. We will now discuss why. Firstly, it is worth being noted that we do not have at hand any theoretical results to compare behaviour of the different selection rules that have been investigated. In fact, theoretical results in regression are very much difficult to get because of the random denominator of Watson-Nadaraya estimate. Excepted for Cross-Validation for which this denominator problem has been solved (lemma 4 in Härdle and Marron 1985, and lemma 5 in Härdle and Vieu 1991), all other methods are studied only in the case when the density of X is known. This is of course very restrictive, and this restriction does not allow us to have theoretical mean to compare all these methods. So, an interesting question that could help for comparison is: **Are Plug-in and Bootstrap methods also working in the random design case?**

In the fixed design case, the rates of convergence in (4.1) and (4.3) seem to indicate that Plug-in rule can be (at least asymptotically) more efficient than

146

Cross-Validation type selectors. But to confirm such an idea these results should be shown for the same model and under similar assumptions. Unfortunately this is not yet the case. In the same spirit, a natural question that is also arrising is the following one: **Is Bootstrap rule also able to reach the same rate as Plug-in method?** Related works in density estimation by Marron (1991b), may let expect for a positive answer.

In addition to statements of results of the kind of (4.1), other interesting directions of researches that would be of great help in understanding differences between bandwidth selection rules, is the study of exact errors for finite sample size (see Marron and Wand (1991) for such a study in density estimation). Also interesting would be to develop comparison of these different methods through several different examples. In the related setting of density estimation, a lot of works exist in this direction (see eg Marron 1988, Park and Marron 1990, and Jones 1990), and it appears that there is no definitive preference for one among these methods.

Let us now mention some related works concerning extension of the basic regression models. For instance, Gonzalez-Manteiga and Cadarso-Suarez (1991) propose a modification of Cross-Validation rule to deal with censored regression. Another modification of Cross-validation is introduced by Boularan et al. (1991) to deal with regression with repeated measurements when X is random. Same approach when X is not random is studied by Wehrly and Hart (1988) and Hart and Wherly (1991). Härdle et al (1992b) discuss bandwidth choice for estimation of derivatives of the regression function, and this is of particular interest for Projection Pursuit Regression. Also concerning curse of dimensionality, Vieu (1991d) shows how cross-validated bandwidth can be used to estimate non-parametrically the order of an autoregressive process.

The techniques used in regression for bandwidth selection are also very common in other functional estimation problems. We cited along this paper several references concerning bandwidth selection in density estimation. In addition the works by Diggle and Marron (1988) in intensity function estimation, by Sarda (1991) in distribution function estimation, and by Sarda and Vieu (1991), Patil (1991) and Müller and Wang (1990) in hazard estimation, are worth being cited. As in regression, most of these papers are dealing with Cross-Validation techniques.

Acknowledgments: The author would like to express his gratitude to Steve Marron. His interesting suggestions have certainly improved this paper. Many thanks also to an anonymous referee.

References

AKAIKE, H. (1970). Statistical predictor identification. *Ann. Inst. Statist. Math.*, **22**, 203-217.

AKAIKE, H. (1974). A new look at the statistical model identification. *I.E.E.E. Trans. Auto. Control*, **19**, 716-723.

ALTMAN, N. (1990). Kernel smoothing of data with correlated errors. *J. Amer. Statist. Assoc.*, **85**, 749-759.

BURMAN, P. and CHEN, K.W. (1989). Nonparametric estimation of regression function. *Ann. Statist.*, **17**, 1567-1596.

BOULARAN, J., FERRE, L. and VIEU, P. (1991). Growth curves: a two-stage nonparametric approach. Preprint.

CAO-ABAD, R. (1991). Rate of convergence for the wild bootstrap in nonparametric regression. *Ann. Statist.*, **19**, 2226-2231.

CHIU, S.T. (1989). Bandwidth selection for kernel estimate with correlated noise. *Statist. Proba. Let.*, **8**, 347-353.

CHIU, S.T. (1990). On the asymptotic distribution of bandwidth estimates. *Ann. of Statist.*, **18**, 1696-1711.

CHIU, C. and MARRON, J.S. (1990). The negative correlation between data determined bandwidth and the optimal bandwidth. *Statist. Proba. Let.*, **10**, 173-180.

CHU, C. and MARRON, J.S. (1991a). Comparison of two bandwidth selectors for dependent errors. *Ann. Statist.*, **19**, 1906-1918.

CHU, C. and MARRON, J.S. (1991b). Choosing a kernel regression estimator. *Statist. Science.*, **6**, 404-436.

COLLOMB, G. (1976). Estimation non paramétrique de la régression par la méthode du noyau. *Thesis at Univ. P. Sabatier, Toulouse, France.*

COLLOMB, G. (1981). Estimation non paramétrique de la régression : revue bibliographique. *Inter. Statist. Review*, **49**, 75-93.

COLLOMB, G. (1985). Nonparametric regression an up-to-date bibliopgraphy. *Statistics*, **2**, 309-324.

COLLOMB, G., SARDA, P. and VIEU, P. (1986). Weak pointwise consistency of the cross cvalidatory window estimate in non parametric regression. *Comm. Math. Univ. Carlova*, **26**, 789-798.

DIGGLE, P. and MARRON, J.S. (1988) Equivalence of smoothing parameter selectors in density and intensity. *J. Amer. Statist. Assoc.*, **83**, 793-800.

EUBANK, R.L. and SCHUCANY, W. (1990). Adaptive bandwidth choice for kernel regression. Preprint.

FAN, J. (1992a). Design adaptive regression. *J. Amer. Statist. Ass.*, **87**, in print.

FAN, J. (1992b). Variable bandwidth and local linear regression smoothers. Preprint.

FAN, J. (1992c). Local linear regression smoothers. Preprint.

GASSER, T. and HERRMANN, E. (1991). Data-adaptive kernel estimation. in *Nonparametric functional Estimation and related topics.* Ed G.G. Roussas. Kluwer Academic Publishers, 67-80.

GASSER, T., KNEIP, A. and KOHLER, W. (1991). A flexible and fast method for automatic smoothing. *J. Amer. Statist. Assoc.*, **86**, 643-653.

GASSER, T. and MÜLLER, H.G. (1984). Nonparametric estimation of regression functions and their derivatives by the kernel method. *Scand. J. of Statist.*, **11**, 171-185.

GOLDSTEIN, L. and MESSER, K. (1990). Optimal plug-in estimators for nonparametric functional estimation. Preprint.

GONZALEZ-MANTEIGA, W. and CADARSO-SUAREZ, M. (1991). A generalized Kaplan Meier estimator with applications. Preprint.

GYÖRFI, L., HÄRDLE, W., SARDA, P. and VIEU, P. (1989). *Nonparametric curve estimation from time series.* Springer, Lectures notes in Statistics, vol 60.

HALL, P. (1984a). Asymptotic properties of integrated squared errors and cross-validation for kernel estimation of a regression function. *Z.f.W.u.v.G.*, **67**, 175-196.

HALL, P. (1984b). Integrated squared errors properties of kernel estimation of a regression function. *Ann. Statist.*, **12**, 241-260.

HALL, P. (1989). On convergence rates in nonparametric problems. *Int. Statist. Rev.*, **57**, 47-58.

HALL, P. and JONES, C. (1990). Adaptive M-estimation in nonparametric regression. *Ann. Statist.*, *B*, **53**,245-252.

HALL, P. and MARRON, J.S. (1991). Local minima in Cross-Validation. *J. Roy. Statist. Soc.*, *B*, **53**,245-252.

HÄRDLE, W. (1990). *Applied nonparametric regression*. Oxford University Press, Boston.

HÄRDLE, W. and BOWMAN, A. (1987). Bootstrapping in nonparametric regression. *J.A.S.A.*, **83**, 127-141.

HÄRDLE, W., HALL, P. and MARRON, J.S. (1988). How far are automatically chosen regression smoothing parameters from their optimum? *J. Amer. Statst. Assoc.*, **83**, 86-101. With discussions.

HÄRDLE, W., HALL, P. and MARRON, J.S. (1992a). Regression smoothing parameters that are not far from their optimum? *J. Amer. Statst. Assoc.*, In print.

HÄRDLE, W., HART, J., MARRON, J.S. and TSYBAKOV, A. (1992b). Bandwidth choice for average derivative estimation. *J. Amer. Statst. Assoc.*, **87**, 218-226.

HÄRDLE, W. and KELLY, G. (1987). Nonparametric kernel regression estimation. Optimal bandwidth choice. *Statistics*, **18**, 21-35.

HÄRDLE, W. and MAMMEN, E. (1990). Bootstrap methods in nonparametric regression. *Proccedings of NATO advanced study on Nonparametric Functional Estimation, Kluwer Dordrecht (ed G.G. ROUSSAS)*, 111-123.

HÄRDLE, W. and MARRON, J.S. (1985a). Optimal bandwidth selection rule in nonparametric regression. *Ann. Statist.* **13**, 4, 1465-1481.

HÄRDLE, W. and MARRON, J.S. (1985b). Asymptotic nonequivalence of some bandwidth selectors in nonparametric regression. *Biometrika*, **72**, 481-484.

HÄRDLE, W. and MARRON, J.S. (1990). Semiparametric comparison of regression curves. *Ann. Statist.* **18**, 63-89.

HÄRDLE, W. and VIEU, P. (1991). Kernel regression smoothing of time series. *J. of Time Series Anal.*, in print.

HART, J. and VIEU, P. (1990). Data-driven bandwidth choice for density estimation based on dependent data. *Ann. Statist.*, **18**, 873-890.

HART, J. and WEHRLY, T, (1991). Consistency of cross-validation when data are curves. Accepted for publication in *Stoch. Proc. Appl.*.

HERRMANN, E., GASSER, T. and KNEIP, A. (1991). Choice of bandwith for kernel regression when residuals are correlated. Preprint.

JOHNSTON, G.J. (1982). Probabilities of maximal deviations for nonparametric regression function estimates. *J. Mult. Anal.*, **12**, 402-444.

JONES, M.C. (1990). Prospects for automatic bandwidth selection in extensions to basic kernel estimation. *Proccedings of NATO advanced study on Nonparametric Functional Estimation, Kluwer Dordrecht (ed G.G. ROUSSAS)*, 241-250.

JONES, M.C. and DAVIES, S. (1991). Versions of kernel regression estimator. Preprint.

KONAKOV, V.D. and PITERBARG, V.I. (1984). On the convergence rate of maximal deviation distribution for kernel regression estimates. *Journ. Mult. Anal.*, **15**, 279-294.

MACK, Y. and WULWICK, M.C. (1991). Nonparametric regression analysis of some economic data. *Proccedings of NATO advanced study on Nonparametric Functional Estimation, Kluwer Dordrecht (ed G.G. ROUSSAS)*, 361-374.

MACK, Y. and MÜLLER, H.G. (1989a). Derivative estimation in nonparametric regression with random predictor variable. *Sankhya*, *A*, **51**, 59-72.

MACK, Y. and MÜLLER, H.G. (1989b). Convolution type estimators for nonparametric regression. *Statist. Proba. Let.*, **7**, 229-239.

MARRON, J.S. (1987). Partitioned cross-validation. *Econometric Reviews*, **6**, 271-284.

MARRON, J.S. (1988). Automatic smoothing parameter selection: a survey. *Empirical Economics*, **13**, 187-208.

MARRON, J.S. (1991). Bootstrap bandwidth selection. Preprint.

MARRON, J.S. and HÄRDLE, W. (1986). Random approximations of some measures of accuracy in nonparametric curve estimation. *J. Multiv. Anal.*, **20**, 91-113.

MARRON, J.S. and WAND, M. (1991). Exact mean integrated squared error. Preprint.

MÜLLER, H.G. (1985). Empirical bandwidth choice for nonparametric regression by means of pilot estimators. *Statist. and Decisions*, **2**, 193-206.

MÜLLER, H.G. and WANG, J.L. (1990). Locally adaptive hazard smoothing. *Prob. Th. Rel. Fields*, **85**, 523-538.

NADARAYA, E. (1964). On estimating regression. *Theory Proba. and Appl.*, **9**, 141-142.

PATIL, P. (1991). Bandwidth choice for non parametric hazard rate estimation. Preprint.

PARK, B. and MARRON, J.S. (1990). Comparison of data-driven bandwidth selectors. *J. Amer. Statist. Ass.*, **85**, 66-72.

PARZEN, E. (1962). On estimating a probability density function and the mode. *Ann. Math. Statist.*, **33**, 1065-1076.

PRIESTLEY, M.B. and CHAO, M.T. (1972). Non-parametric function fitting. *J. Roy. Statist. Ass., ser. B*, **34**, 385-392.

RICE, J. (1984). Bandwidth choice for nonparametric regression. *Ann. Statist.*, **33**, 1215-1230.

SARDA, P. (1991). Estimating smooth distribution functions. *Proceedings of NATO advanced study on Nonparametric Functional Estimation, Kluwer Dordrecht (ed G.G. ROUSSAS)*, 261-270.

SARDA, P. and VIEU, P. (1991). Smoothing parameter selection in hazard estimation. *Statist. Proba. Let.*, **11**, 429-434.

SHIBATA, R. (1981). An optimal selection of regression variables. *Biometrika*, **68**, 45-54.

STANISWALLIS, J.G. (1989). Local bandwidth selection for kernel estimate. *J. Amer. Statist. Ass.*, **84**.

STONE, C. (1982). Optimal global rates of convergence in nonparametric regression. *Ann. Statist.*, **10**, 1040-1053.

VIEU, P. (1991a). Quadratic errors for nonparametric estimates under dependence. *J. Multivariate Anal.*, **39**, 324-347.

VIEU, P. (1991b). Smoothing techniques in Time Series Analysis. *Proceedings of NATO advanced study on Nonparametric Functional Estimation, Kluwer Dordrecht (ed G.G. ROUSSAS)*, 271-283.

VIEU, P. (1991c). Nonparametric regression: local optimal bandwidth choice. *J. Roy. Statist. Soc. B*, **53**, 453-464.

VIEU, P. (1991d). On nonparametric autoregression order choice. Preprint.

WATSON, G.S. (1964). Smooth regression analysis. *Sankhya, A*, **26**, 359-372.

WEHRLY, T, and HART, J. (1988). Bandwidth selectors for kernel estimators of growth curves with correlated errors (abstract). *Inst. Math. Statist. Bull.*, **17**.

WONG, W. (1982). On the consistency of cross validation in kernel non-parametric regression. *Ann. Statist.*, **11**, 1136-1141.

Practical Use of Bootstrap in Regression

Marie-Anne GRUET, Sylvie HUET and Emmanuel JOLIVET

INRA, Laboratoire de Biométrie, F78352 JOUY-EN-JOSAS Cédex, France

Abstract :

The usefulness of bootstrap in statistical analysis of regression models is demonstrated. Surveying earlier results, four specific problems are considered:

the computation of confidence intervals for parameters in a nonlinear regression model,

the computation of calibration sets in calibration analysis, when the standard curve is described by a nonlinear function,

the estimation of the covariance matrix of the parameter estimates for an incomplete analysis of variance model, in the presence of an interaction term,

the computation of confidence intervals for the value of the regression function, when a nonparametric heteroscedastic model is considered.

Theoretical properties of the proposed bootstrap procedures, as well as indications about their actual efficiency based on simulation results, are given.

Key words: bootstrap, calibration, analysis of variance, nonlinear regression, nonparametric regression, Edgeworth expansion, bootstrap

1 Introduction

For many statistical models of frequent practical use, only the asymptotic theory is available. This is the case for the regression models, except for completely parametric models, where the regression function is linear with respect to the parameters, and the distribution of the errors is Gaussian. The applied statistician is thus faced with the following question: "How much may I rely on the theoretical asymptotic results when I'm performing an actual analysis on real data, most of the time in a *small sample* situation?". Next, he's asking for methods, improving the crude asymptotic first order methods, straightforward consequences of the law of large numbers and of the central limit theorem, in the framework of models as general as possible.

The resampling procedures, and especially the bootstrap, are among the promising methods to solve practically that problem. Their success is mainly due to the very good approximation of a distribution function by the corresponding empirical process, as stated by the Glivenko-Cantelli theorem. For instance, in many situations, the bootstrap achieves a *second order correct* approximation of the distribution functions of statistical quantities of interest, as asymptotic pivots. And the price to pay is only the one of some repetitive, although sometimes overwhelmingly intensive, computations. This fact was first stressed by Singh (1981) for the Studentised mean in the i.i.d. situation, and broadly extended, mainly by Beran (1987) and Hall (1986), Hall (1988), to more general statistical objects.

In the present survey paper, we summarize results previously established concerning the use of bootstrap procedures in analysing regression models. The main purpose is to construct confidence sets for real quantities of interest, and to control as well as possible the level of these confidence sets. Four different problems are considered. First, we consider the computation of confidence intervals for the parameters of the regression function in a nonlinear regression model. Second, calibration sets are constructed, in the case of a nonlinear parametric function as a model for the standard curve. In both above situations, the basic statistical model is a semiparametric one, as we don't assume a definite parametric family for the distribution function of the errors. On the other hand, the third model considered is a fully parametric one. It's a two factors analysis of variance model, with a multiplicative interaction term, and Gaussian errors. Incomplete designs, implying difficult computations for estimating the covariance matrix of the parameters estimate, are considered. Finally, the fourth problem concerns a nonparametric heteroscedastic regression model, and confidence intervals for the value of the regression function at a fixed point.

In each case, the bootstrap furnishes relevant tools to the applied statisticians faced with these problems. The relevance is proved theoretically by asymptotic arguments for all the four situations. For the first two problems, simulation results suggest good actual performances of these resampling methods for a small or moderate sample size. A real-size genotype-location assay has been analyzed using a method arising from the study of the third problem.

These results are reviewed in the sequel, after a short paragraph on the main methodological features shared between the four problems. No formal proofs are given, but complete references to the original papers are indicated.

2 Why does the bootstrap work?

Although the grounds for the good behaviour of the bootstrap are well known today, let's try to present them briefly in a comprehensive and consistent way, embracing the needs of our four problems.

Firstly, it's obvious that the only theoretical justification of the bootstrap method is obtained via asymptotic results. Actually, consider the scalar random variable of interest T calculated from an observation $\mathcal{X}$ of size n, call T^* the bootstrap version of T, obtained after a convenient resampling scheme based on $\mathcal{X}$. In the various situations considered in details hereafter. the key-result is of the following form:

$$\lim_{n \to \infty} \sup_{t \in \mathbf{R}} \alpha_n \left| \Pr\{T \le t\} - \Pr\{T^* \le t | \mathcal{X}\} \right| = 0 \,, \tag{1}$$

for almost all observation $\mathcal{X}$. Here, α_n is some real quantity tending to infinity with the sample size. Thus, it's a good measure of the performance of the bootstrap.

Secondly, the way to determine the proper order of magnitude of α_n is by expanding the distribution function of T and the conditional distribution function of T^*. A lot of work has been dedicated to the validity of the Edgeworth expansion (Bhattacharya and Ghosh (1978), Hall (1983) for instance). Assume that the distribution function of T converges to $\Psi(.)$, the distribution function of some random variable, not depending on the parameters of the statistical model for $\mathcal{X}$. Let $\psi(.)$ be the derivative of $\Psi(.)$. Under convenient regularity conditions, results of the following pattern are obtained:

$$\lim_{n \to \infty} \sup_{t \in \mathbf{R}} \alpha_n' \left| \Pr\{T \le t\} - \Psi(t) - \frac{1}{\delta_n^{(1)}} \psi(t) P_{(1)}(t) - \ldots - \frac{1}{\delta_n^{(r)}} \psi(t) P_{(r)}(t) \right| = 0 \,, \tag{2}$$

where $\delta_n^{(j)}$, $j = 1, \ldots, r$ are quantities tending to infinity with n, and with an order of magnitude increasing with j, α_n' tending also to infinity with n, at least at the same rate as $\delta_n^{(r)}$. The functions $P_{(1)}(.), \ldots, P_{(r)}(.)$ are polynomials depending on the moments of the observations distribution function. In many situations, Ψ is the distribution function of a standard Gaussian variable, noted Φ, and ψ its density, noted ϕ.

The same type of result is true for the conditional distribution function of T^*, if the above mentioned regularity conditions are also fulfilled in the *bootstrap world*. But the statistical model is essentially the same in both the real and the bootstrap world, except for the distribution function of the observations. Henceforth, if a nonparametric bootstrap procedure is considered, the crucial point is to prove that the so called *Cramer condition*, concerning the characteristic function of (a function of) the observations is also verified for the characteristic function of the (corresponding function of the) pseudo-observations generated by the resampling procedure. Note that this point is often omited in the literature, although it's not a trivial one. The key is given by a theorem of Csörgő and Totik (1983) on the accuracy of the approximation of the characteristic function by its empirical counterpart. When the observations to be resampled aren't directly observed, but are to be estimated, this result has to be adapted. It's the case in a regression analysis, where often the residuals, and not the errors, are resampled. For the nonlinear regression, such a result (see Huet and Jolivet (1989)) has been established.

Thus, assuming the abovementioned regularity conditions to be fulfilled, a paraphrase of equation (2) is obtained:

$$\lim_{n\to\infty} \sup_{t\in\mathbf{R}} \alpha'_n \left| \Pr\{T^* \leq t \,|\, \mathcal{X}\} - \Psi(t) - \frac{1}{\delta_n^{(1)}} \psi(t) \hat{P}_{(1)}(t) - \ldots - \frac{1}{\delta_n^{(r)}} \psi(t) \hat{P}_{(r)}(t) \right| = 0 \,, \qquad (3)$$

this result being true almost surely. Here, the functions $\hat{P}_{(1)}(.), \ldots, \hat{P}_{(r)}(.)$ are the same as the polynomials $P_{(1)}(.), \ldots, P_{(r)}(.)$, but where the moments of the distribution function of the pseudo observations generated by the resampling scheme stand for the true moments. As such, the coefficients of the $\hat{P}$ can be viewed as estimates of the P based on the observation $\mathcal{X}$.

Finally, assume that $\max_{j=1,\ldots,r} \sup_{t\in\mathbf{R}} \left| \hat{P}_{(j)}(t) - P_{(j)}(t) \right| = O_p(\beta_n^{-1})$, where β_n tends to infinity with n. If the order of magnitude of $\delta_n^{(1)}\beta_n$ is greater than the order of magnitude of $\delta_n^{(k+1)}$, then, roughly speaking, the expansions of the distribution function of T and of the conditional distribution function of T^* agree up to the order of magnitude of $\delta_n^{(k)}$. More precisely, this result is expressed by the equation (1), with $\alpha_n = \delta_n^{(k)}$.

After recalling the main features of the theoretical machinery, let's consider in more details the four announced problems.

3 Confidence intervals for the parameters of a semiparametric nonlinear regression model

3.1 Theoretical background

Assume that $\mathcal{X} = X_1, \ldots, X_n$ are observed, with

$$X_i = f(x_i, \theta_o) + \varepsilon_i \,, \quad i = 1, \ldots, n \,, \qquad (4)$$

where $x_1, \ldots, x_n$ is a set of known real numbers, f is a known real function depending nonlinearly on an unknown parameter θ, which belongs to a compact subset Θ of $\mathbf{R}^p$. The ε_i, $i = 1, \ldots, n$ are independent identically distributed real random variables, with expectation 0 and common variance σ^2. Under convenient regularity conditions, it's well known (see Jennrich (1969)) that the least squares estimators of θ_o and σ^2:

$$\hat{\theta}_n = \arg\min_{\theta\in\Theta} \sum_{i=1}^{n} (X_i - f(x_i, \theta))^2$$

and

$$\hat{\sigma}_n^2 = n^{-1} \sum_{i=1}^{n} \left(X_i - f(x_i, \hat{\theta}_n) \right)^2$$

are consistent, and fulfill the weak convergence result

$$\sqrt{n}\hat{\sigma}_n^{-1}\Gamma_{n\hat{\theta}_n}^{1/2} \left(\hat{\theta}_n - \theta_o \right) \xrightarrow{\mathcal{D}} \mathcal{N}(0, I) \qquad (5)$$

where I is the $p \times p$ identity matrix and

$$\Gamma_{n\,\theta} = n^{-1} \left(\sum_{i=1}^{n} \frac{\partial f}{\partial \theta_a}(x_i, \theta) \frac{\partial f}{\partial \theta_b}(x_i, \theta) \right) \quad \begin{array}{l} a = 1, \ldots, p \\ b = 1, \ldots, p \end{array}$$

Moreover, Schmidt and Zwanzig (1986) have given the conditions of the existence of an Edgeworth expansion for

$$T_n = \sqrt{n}\hat{\sigma}_n^{-1}\lambda_{n\,\hat{\theta}_n}^{-1} (\hat{\gamma}_n - \gamma_o) \ ,$$

where γ_o is the a-th coordinate of θ_o, $\hat{\gamma}_n$ is the a-th coordinate of $\hat{\theta}_n$, and $\lambda_{n\,\hat{\theta}_n}^2$ is the a-th element of the diagonal of the matrix $\Gamma_{n\,\hat{\theta}_n}^{-1}$.

Let's consider the following resampling scheme, which is a nonparametric bootstrap, with an estimated empirical distribution function used to generate the observations. Let $\tilde{\varepsilon}_i = \hat{\varepsilon}_i - n^{-1}\sum_{j=1}^{n}\hat{\varepsilon}_j$, where $\hat{\varepsilon}_i = X_i - f(x_i, \hat{\theta}_n)$, and let $\tilde{F}_n$ denote the empirical distribution function of these centered residuals:

$$\tilde{F}_n(t) = n^{-1}\sum_{i=1}^{n} 1_{\{\tilde{\varepsilon}_i \leq t\}} \ .$$

Then, the following pseudo-observations are generated:

$$X_i^* = f(x_i, \hat{\theta}_n) + \xi_i \ , \quad i = 1, \ldots, n \ ,$$

where $\xi_1, \ldots, \xi_n$ is an i.i.d. sample with the distribution function $\tilde{F}_n$. Note that $\tilde{F}_n$ is close to the empirical process based on the ε_i but isn't this empirical process itself, and that it depends on $\hat{\theta}_n$. Let's call $\hat{\theta}_n^*$ and $\hat{\sigma}_n^{*\,2}$ the least squares estimators of θ_o and σ^2 obtained from the pseudo observations X_i^*, $i = 1, \ldots, n$, and define T_n^*

$$T_n^* = \sqrt{n}\hat{\sigma}_n^{*\,-1}\lambda_{n\,\hat{\theta}_n^*}^{-1} (\hat{\gamma}_n^* - \hat{\gamma}_n) \ ,$$

where $\lambda_{n\,\hat{\theta}_n^*}$ and $\hat{\gamma}_n^*$ have a straightforward meaning. Huet and Jolivet (1989) gave regularity conditions for the existence of an Edgeworth expansion for the conditional distribution function of T_n^*, under which the following statement is true:

$$\lim_{n\to\infty} \sup_{t\in\mathbf{R}} \sqrt{n} \left| \Pr\{T_n \leq t\} - \Pr\{T_n^* \leq t \,|\, \mathcal{X}\} \right| = 0 \tag{6}$$

almost surely.

The approximation of the distribution function of T_n obtained by the resampling scheme just described is then exact up to the order $n^{-1/2}$, and as such is better than the approximation given by the equation (5).

3.2   Practical implementation

The preceding results provide various methods for computing confidence intervals for γ_o based on the distribution of the asymptotic pivot T_n. Only three of them are considered here. The first one is based on the crude application of the equation (5) and is given by:

$$I_N = \left[\hat{\theta}_n - n^{-1/2} \hat{\sigma}_n \lambda_{n\hat{\theta}_n} u_{1-\alpha}, \ \hat{\theta}_n - n^{-1/2} \hat{\sigma}_n \lambda_{n\hat{\theta}_n} u_\alpha \right] ,$$

where u_α is the α-quantile of the standard Gaussian distribution. The second one is a slight extension of the preceding, obtained by a simple transposition of the exact results obtained in the framework of the linear Gaussian regression:

$$I_S = \left[\hat{\theta}_n - (n-p)^{-1/2} \hat{\sigma}_n \lambda_{n\hat{\theta}_n} t_{n-p\,1-\alpha}, \ \hat{\theta}_n - (n-p)^{-1/2} \hat{\sigma}_n \lambda_{n\hat{\theta}_n} t_{n-p\,\alpha} \right] ,$$

where $t_{n-p\,\alpha}$ is the α-quantile of the Student distribution with $n - p$ degrees of freedom. The third one is based on the quantiles of the conditional distribution of T_n^*. Let $\mathsf{b}_\alpha(\mathcal{X}) = \inf \{ t : \Pr \{ T_n^* \le t \,|\, \mathcal{X} \} \ge \alpha \}$. It gives rise to the bootstrap confidence interval

$$I_B = \left[\hat{\theta}_n - n^{-1/2} \hat{\sigma}_n \lambda_{n\hat{\theta}_n} \mathsf{b}_{1-\alpha}(\mathcal{X}), \ \hat{\theta}_n - n^{-1/2} \hat{\sigma}_n \lambda_{n\hat{\theta}_n} \mathsf{b}_\alpha(\mathcal{X}) \right] .$$

A simulation study, including these confidence intervals as well as others, has been performed and is reported in Huet *et al.* (1989). Some of the outcomes are quoted here.

Let's consider the *Bleasdale-Nelder* model, with

$$f(x, \theta) = -\frac{1}{\theta_3} \log(\theta_1 + \theta_2 x) ,$$

where the parameters are given the following values

$$\theta_1 = 0.001897 , \quad \theta_2 = 0.0001095 , \quad \theta_3 = 1.0 , \quad \sigma^2 = 0.15 , \quad n = 20 . \quad x_i = 10i , i = 1, \ldots . 20 .$$

and the *exponential* model, with

$$f(x, \theta) = \theta_1 + \theta_2 \exp(\theta_3 x) ,$$

where the parameters are given the following values

$$\theta_1 = 1122.2 , \quad \theta_2 = -1308.7 , \quad \theta_3 = -0.08715 , \quad \sigma^2 = 40.0 , \quad n = 15 , \quad x_i = i , i = 1, \ldots , 15 .$$

For each model, $N = 500$ simulations have been carried out, that is N times n observations $X_1, \ldots, X_n$ have been generated, with $X_i = f(x_i, \theta) + \varepsilon_i$, ε_i being given by a (pseudo) random number generator generating independent centered normal variates with variance σ^2. For each simulation, the least squares procedure has been applied, and $B = 500$ bootstrap simulations have been performed according to the resampling scheme described above. Some results are displayed on the tables (1) and (2). They are given for two values of α: 0.25 and 0.025. The column labeled $\{\gamma \in I\}$ contains the proportion of confidence intervals containing the true value of the parameter. the column labeled $\{\gamma < \underline{I}\}$ contains the proportion of confidence intervals with the true value of the parameter on the left hand side of its lower bound, the column labeled $\{\gamma > \overline{I}\}$ contains the proportion of confidence intervals with the true value of the parameter on the right hand side of its upper bound.

		$\{\gamma \in I\}$	$\{\gamma < \underline{I}\}$	$\{\gamma > \bar{I}\}$
θ_1 50%	I_N	57.7%	10.5%	31.8%
	I_S	63.6%	5.8%	30.6%
	I_B	50.7%	23.9 %	25.3%
θ_1 95%	I_N	82.5%	0%	17.5%
	I_S	85.1%	0%	14.9%
	I_B	94.6%	0%	5.4%
θ_3 50%	I_N	46.1%	25.7%	28.2%
	I_S	51.9%	23.3%	24.7%
	I_B	48.5%	27.2%	24.3%
θ_3 95%	I_N	92.1%	4%	3.8%
	I_S	95.5%	2.4%	2%
	I_B	91.9%	6%	2%

Table 1: Bleasdale-Nelder model: comparison of coverage probabilities for three different confidence intervals and two asymptotic levels

		$\{\gamma \in I\}$	$\{\gamma < \underline{I}\}$	$\{\gamma > \bar{I}\}$
θ_1 50%	I_N	46.6%	25.6%	27.8%
	I_S	53%	21.6%	25.4%
	I_B	49.6%	27.6%	22.8%
θ_1 95%	I_N	91.2%	0.4%	8.4%
	I_S	94.6%	0%	5.4%
	I_B	89.8%	8.2%	2%
θ_3 50%	I_N	45.6%	29.2%	25.2%
	I_S	51.6%	26%	22.4%
	I_B	50.8%	26.8%	22.4%
θ_3 95%	I_N	92.2%	5.2%	2.6%
	I_S	97%	1.8%	1.2%
	I_B	96%	3%	1%

Table 2: Exponential model: comparison of coverage probabilities for three different confidence intervals and two asymptotic levels

4 Calibration sets, when the standard curve is described by a nonlinear regression model

Assume z_0 to be an unknown scalar quantity to be measured, but impossible to observe directly. Another scalar quantity Z is observed, such that

$$Z = f(z_0, \theta_0) + \eta$$

where η is a centered random variable with expectation 0 and finite variance v^2. f and θ_0 are as in the preceding section. In order to make any inference, it's necessary to get an additional information on the link between z_0 and Z. This is achieved by taking a series of known values $x_1, \ldots, x_n$, and observing $X_1, \ldots, X_n$, the statistical model being given by equation (4), with the same hypothesis. So the least squares procedure given in section 3 can be used to estimate θ_0 and σ^2, and the same properties are fulfilled. In addition, f is assumed to be strictly monotone, and Z independent of the X_i. In such a framework, the graph of the equation $y = f(x, \theta_0)$ is called the standard curve. The statistical properties of any estimator of z_0 depend on the variability induced by the observations used to estimate the standard curve, as well as by Z. In practical studies, the user is often more interested by a set estimation than by a point estimation. We review briefly results by Gruet and Jolivet (1991) along these lines.

4.1 Theoretical background

Let's begin by an obvious remark: $Z - f(z_0, \hat{\theta}_n) = \eta - \left(f(z_0, \hat{\theta}_n) - f(z_0, \theta_0) \right)$ is the sum of two independent quantities. Thus, if their distributions are known, or if the law of derived statistical quantities, as pivots, are known, a straightforward way to construct a confidence set for z_0 is open. This is the situation if the model for the standard curve is a linear regression with Gaussian errors.

Let's also remark that $\hat{z} = f^{-1}(Z, \hat{\theta}_n)$ is the least squares estimator of z_0. The distribution function of (a function of) $\hat{z} - z_0$, if it's known, provides another way to construct a confidence interval. Again, in the linear Gaussian situation, an exact confidence interval can be obtained. But, in a general framework, some information on the distribution of Z is to be known. It's assumed here to be obtained through some replications:

$$\mathcal{Z} = \{ Z_j = f(z_0, \theta_0) + \eta_j , \ j = 1, \ldots, m \} .$$

Let's call $\bar{Z}_m$ the mean of the Z_j and $\hat{v}_m^2$ their empirical variance. The variance of $\bar{Z}_m - f(z_0, \hat{\theta}_n)$ is estimated by

$$\widehat{W}_{m,n}^2(z_0) = m^{-1}\hat{v}_m^2 + n^{-1}\hat{\sigma}_n^2 f_\theta(z_0, \hat{\theta}_n)^T \Gamma_{n \, \hat{\theta}_n}^{-1} f_\theta(z_0, \hat{\theta}_n) ,$$

where f_θ is the vector of the partial derivatives of f with respect to the components of θ. Under convenient regularity conditions, the following weak convergence result follows:

$$S_{m,n} = \widehat{W}_{m,n}^{-1}(z_o) \left(\bar{Z}_m - f(z_o, \hat{\theta})_n \right) \xrightarrow{\mathcal{D}} \mathcal{N}(0,1) \tag{7}$$

when both m and n tend to infinity.

On the other hand, if we assume $\hat{z}_{m,n} - z_o$, with $\hat{z}_{m,n} = f^{-1}(\bar{Z}_m, \hat{\theta}_n)$, sufficiently small. its variance can be estimated by

$$\widehat{V}_{m,n}^2 = m^{-1}\hat{v}_m^2 \left(f_z^{-1}(\bar{Z}_m, \hat{\theta}_n) \right)^2 + n^{-1}\hat{\sigma}_n^2 f_\theta^{-1}(\bar{Z}_m, \hat{\theta}_n)^T \Gamma_{n \hat{\theta}_n}^{-1} f_\theta^{-1}(\bar{Z}_m, \hat{\theta}_n) ,$$

where f_z^{-1} (resp. f_θ^{-1}) denotes the partial derivative of f^{-1} with respect to z (resp. the vector of partial derivatives of f^{-1} with respect to the components of θ). Another weak convergence result follows, again under regularity hypothesis:

$$T_{m,n} = \widehat{V}_{m,n}^{-1} \left(\hat{z}_{m,n} - z_o \right) \xrightarrow{\mathcal{D}} \mathcal{N}(0,1) \tag{8}$$

as m and n tend to infinity.

Obviously, $S_{m,n}$ and $T_{m,n}$ are the same if f is a linear function of the parameters. But in a nonlinear regression framework, these asymptotic pivots are different and it's worth comparing them, through the Edgeworth expansions of their distribution functions for instance.

However, the characteristics of the asymptotic have to be defined first. In most practical situations, n is typically of the order of some tens and m of some unities. If m is held fixed whereas n tends to infinity, only the variability of $\bar{Z}_m$ is to be taken into account, $\hat{\theta}_n$ being considered as a perfect estimator of θ_o. If m and n tend to infinity at the same rate, a good representation of actual designs of experiment is certainly not achieved. By choosing a situation where m and n tend to infinity, but with m of the same order as n^δ, $0 < \delta < 1$, it's intended to give a relevant insight into the actual behaviour of $T_{m,n}$ and $S_{m,n}$ at finite distance. By the way, indications about the interplay between the variability of $\bar{Z}_m$ and the variability of $\hat{\theta}_n$ are expected.

In Gruet and Jolivet (1991), existence conditions for an Edgeworth expansion for $S_{m,n}$ and $T_{m,n}$ are given. The terms of order m^{-1} are shown to differ and to be in favour of $S_{m,n}$ if the distribution of η is symmetrical. A nonparametric bootstrap resampling procedure is also proposed. The resampling scheme of the pseudo-observations $X_1^*, \ldots, X_n^*$ used to estimate the standard curve is the same as the one of the preceding section. Moreover. pseudo-observations $Z_1^*, \ldots, Z_m^*$ are generated according to the empirical process of the $Z_j, j = 1, \ldots, m$. With plain notations, the bootstrap counterparts of $S_{m,n}$ and $T_{m,n}$ are given by $S_{m,n}^* = \widehat{W}_{m,n}^{*-1}(\hat{z}_{m,n}) \left(\bar{Z}_m^* - f(\hat{z}_{m,n}, \hat{\theta})_n^* \right)$ and $T_{m,n}^* = \widehat{V}_{m,n}^{*-1} \left(\hat{z}_{m,n}^* - \hat{z}_{m,n} \right)$. If δ is restricted to some range (in fact $]2/3, 4/5]$, see the above-mentioned reference for explanation), it turns out that, under suitable regularity conditions

$$\lim_{m \to \infty} \sup_{t \in \mathbf{R}} \frac{n}{\sqrt{m}} \left| \Pr\left\{ S_{m,n}^* \leq t \,|\, \mathcal{X}, \mathcal{Z} \right\} - \Pr\left\{ S_{m,n} \leq t \right\} \right| = 0$$

and

$$\lim_{m\to\infty} \sup_{t\in\mathbf{R}} \frac{n}{\sqrt{m}} \left| \Pr\left\{ T^*_{m,n} \le t \,|\, \mathcal{X}, \mathcal{Z} \right\} - \Pr\left\{ T_{m,n} \le t \right\} \right| = 0 \,.$$

4.2 Practical implementation

Using the preceding results, four calibration sets are constructed. The first one, I_R is based on $T_{m,n}$, the second one, I_P, is based on $S_{m,n}$. The third and fourth one, $I_R^{(b)}$ and $I_P^{(b)}$, are the bootstrap versions of I_R and I_P respectively. More precisely,

$$I_R = \left[\hat{z}_{m,n} - u_{1-\alpha}\hat{V}_{m,n}, \hat{z}_{m,n} - u_\alpha\hat{V}_{m,n} \right] \,, \tag{9}$$

$$I_P = \left\{ z : \left| \bar{Z} - f(z, \hat{\theta}_n) \right| \le u_{1-\alpha}\widehat{W}_{m,n}(z) \right\} \,, \tag{10}$$

$$I_R^{(b)} = \left[\hat{z}_{m,n} - \hat{V}_{m,n}\mathbf{t}_{1-\alpha}(\mathcal{X}, \mathcal{Z}), \hat{z}_{m,n} - \hat{V}_{m,n}\mathbf{t}_\alpha(\mathcal{X}, \mathcal{Z}) \right] \,, \tag{11}$$

$$I_P^{(b)} = \left\{ z : \widehat{W}_{m,n}(z)\mathbf{s}_\alpha(\mathcal{X}, \mathcal{Z}) \le \check{Z}_m - f(z, \hat{\theta}_n) \le \widehat{W}_{m,n}(z)\mathbf{s}_{1-\alpha}(\mathcal{X}, \mathcal{Z}) \right\} \tag{12}$$

with $\mathbf{t}_\alpha(\mathcal{X}, \mathcal{Z}) = \inf\{t : \Pr\{T^* \le t \,|\, \mathcal{X}\mathcal{Z}\} \ge \alpha\}$ and $\mathbf{s}_\alpha(\mathcal{X}, \mathcal{Z}) = \inf\{t : \Pr\{S^* \le t \,|\, \mathcal{X}\mathcal{Z}\} \ge \alpha\}$. Let's consider the *three parameters logistic model*, with

$$f(x, \theta) = \frac{\theta_1}{1 + \exp(\theta_2 + \theta_3 x)} \,,$$

where the parameters are given the following values

$$\theta_1 = 500 \,, \quad \theta_2 = -5 \,, \quad \theta_3 = 10 \,, \quad \sigma^2 = 100 \,, \quad n = 11 \,, \quad x_i = 0.1(i - 1) \,, i = 1, \ldots, 11 \,,$$

$$z_o = 7/12 \,, \quad m = 4 \,,$$

with homoscedastic Gaussian errors, and the *five parameters logistic model*, with

$$f(x, \theta) = \theta_2 + \frac{\theta_1 - \theta_2}{(1 + \exp(\theta_3 + \theta_4 x))^{\theta_5}} \,,$$

where the parameters are given the following values

$$\theta_1 = 2758.73 \,, \quad \theta_2 = 133.42 \,, \quad \theta_3 = 3.20 \,, \quad \theta_4 = 3.26 \,, \quad \theta_5 = 0.61 \,, \quad \sigma^2 = 0.00087 \,, \quad n = 64 \,.$$

The design of experiment for the standard curve can be found in the abovementioned reference. The errors are distributed according to a Gaussian distribution, with a standard error equal to $\sigma f(x, \theta)$. The resampling scheme is adapted in a straightforward manner, to cope with the variance model. Two values of m (2 and 4), as well as for z_o (-1.25 and -0.7) are used.

For each model, and for each situation, 500 simulations are performed. The size of the bootstrap simulations is 199. The results are given in table (3) and (4).

confidence interval	$\mathcal{J}_R$	$\mathcal{J}_P$	$\mathcal{J}_R^{(b)}$	$\mathcal{J}_P^{(b)}$
coverage probability	0.796	0.802	0.970	0.970

Table 3: Coverage probabilities from the simulation study based on the three parameters logistic model

	$z_0 = -1.25$		$z_0 = -0.7$	
	$m = 2$	$m = 4$	$m = 2$	$m = 4$
$\mathcal{J}_R$	0.744	0.888	0.716	0.894
$\mathcal{J}_P$	0.742	0.890	0.716	0.894
$\mathcal{J}_R^{(b)}$	0.954	0.988	0.946	0.974
$\mathcal{J}_P^{(b)}$	0.954	0.988	0.946	0.976

Table 4: Coverage probabilities from the simulation study on the five parameters logistic model

5 Covariance matrix estimation for an incomplete analysis of variance design

The genotype-location interaction is an important research problem for agronomists and plant breeders. Roughly speaking, the yield of a given plant genotype g at location l is described by the following analysis of variance model

$$X_{gl} = \mu + \alpha_g + \beta_l + \iota_{gl} + \varepsilon_{gl} \tag{13}$$

where g varies in $\{1, \ldots, G\}$, the set of the given genotypes, l varies in $\{1, \ldots, L\}$, the set of the assayed locations, the various parameters being interpreted as usual in an analysis of variance situation. The errors are assumed to be independent, with a common normal distribution, expectation 0 and variance σ^2. As only one observation per cell is assumed to be made, a multiplicative form of the interaction factor is postulated, that is to say $\iota_{gl} = \rho \tau_g \delta_l$. Additional constraints are imposed, to ensure the estimability of the parameters : $\sum_{g=1}^{G} \alpha_g = \sum_{g=1}^{G} \tau_g = \sum_{l=1}^{L} \beta_l = \sum_{l=1}^{L} \delta_l = 0$ and $\sum_{g=1}^{G} \tau_g^2 = \sum_{l=1}^{L} \delta_l^2 = 1$. If the design of experiment is complete, that is to say if each genotype is assayed at each location, explicit calculations of the maximum likelihood estimators of the parameters as well as of their asymptotic covariance matrix are available, the asymptotic situation being described by $\sigma \to 0$ (Chadœuf and Denis (1988), Goodman and Haberman (1990)). But in practice, such a complete design is seldom realised, as new genotypes are introduced, whereas old ones are removed from the experimental scheme. When the design is incomplete, these nice computational properties are no more at hand. However, the model (13) is a particular case of a nonlinear regression, and usual methods of maximum likelihood estimation along the lines described in section 3 can be used. Let's call $\hat{\theta}$ and $\hat{\sigma}^2$ the maximum likelihood estimators of θ_o and σ^2. The correspondence between the regression function of models (4) and (13) is obvious: the parameter θ_o is in some subset Θ of $\mathbf{R}^p$, with $p = 2(L + G) - 4$, taking the constraints on the analysis of variance model into account, and the x_i are the say d indicator functions of the factor combinations assayed in the design of experiment.

By the way, let's remark that the number of observations is fixed in the present case, and that an asymptotic theory based on an infinitely increasing number of observations is not relevant here. The results concerning the properties of the least squares estimators given in section (3) have to be adapted to a variance tending to 0 (Huet (1991)). Under convenient regularity conditions, $T = \sqrt{d - p} \, \hat{\sigma}^{-1} \lambda_{d\hat{\theta}}^{-1} (\hat{\gamma} - \gamma)$ tends in distribution to a Student variable with $d - p$ degrees of freedom. More precisely, St_{d-p} being the distribution function of such a variable,

$$\lim_{\sigma \to 0} \sup_{t \in \mathbf{R}} \sigma^{-1} |\Pr \{T \le t\} - St_{d-p}(t)| < \infty . \tag{14}$$

The notations are selfunderstandable from the section 3.

Nevertheless, the estimation of the parameters asymptotic covariance matrix needs the inversion of the information matrix $\Gamma_{d\hat{\theta}}$. That operation often turns out to be computationally long,

162

difficult, and even unfeasible. Thus, it's tempting to substitute the "brute force" of the bootstrap to that numeric inversion. The complete set of hypothesis made for our analysis of variance model is used, and a parametric resampling scheme is adopted. The pseudo-observations follow the following model:

$$X_{gl}^* = \widehat{\mu} + \widehat{\alpha}_g + \widehat{\beta}_l + \widehat{\rho}\widehat{\tau}_g\widehat{\delta}_l + \varepsilon_{gl}^*$$

where the ε_{gl}^* are iid Gaussian variables with expectation 0 and variance $\widehat{\sigma}^2$, and the pairs (g,l) cover the set of the factor combinations assayed in the design of experiment. For each set of pseudo-observations, bootstrap versions of the estimators of θ and σ^2 are obtained. Performing B bootstrap simulations, B values of a bootstrap estimate for γ, $\gamma^{*(b)}$, $b = 1, \ldots, B$ are obtained and a bootstrap estimate of the variance of $\widehat{\gamma}$ is easily computed. Let $\widehat{V}^{*2} = B^{-1}\sum_{b=1}^{B}\left(\gamma^{*(b)} - B^{-1}\sum_{b=1}^{B}\gamma^{*(b)}\right)^2$ and $T' = \widehat{V}^{*-1}(\widehat{\gamma} - \gamma)$. It can be proved that, if $B^{-1} = o(\sigma)$,

$$\limsup_{\sigma \to 0} \sigma^{-1} |\Pr\{T \leq t\} - \Pr\{T' \leq t | \mathcal{X}\}| = 0$$

almost surely.

On the other hand, a bootstrap version, say T^*, of T can be computed, using the parametric resampling scheme just described above. It can be proved that

$$\limsup_{\sigma \to 0} \sigma^{-1} |\Pr\{T \leq t\} - \Pr\{T^* \leq t | \mathcal{X}\}| < \infty .$$

The rate of convergence is the same than the one obtained in equation (14). In contrast to the results obtained in section 3 and 4, nothing is gained, in terms of rate of convergence, when the distribution function of T is approximated by the conditional distribution function of T^*. Actually, a slight improvement is possibly obtained if $d - p$ is rather big, see Huet (1991) for details. Moreover, a nonparametric resampling design can't be used, because all the results rest heavily on the distributional properties of the observations.

In this rather particular context, the practical interest of the bootstrap is not to provide a method of approximation of the distribution function of T improving the one given by the first order asymptotic, but to make the computations easier, without loss of accuracy up to the order of σ. First attempts of computation of confidence intervals for the parameters of different genotypes of colza assayed on various locations have been performed using the statistic T' and approximating its distribution by a Student, leading to promising preliminary results. They'll be published elsewhere, when completed.

6 Heteroscedastic nonparametric regression

Let's now consider a completely nonparametric framework. n pairs of real numbers (X_i, Y_i), $i = 1, \ldots, n$ are observed, noted $\mathcal{X}, \mathcal{Y}$ for short. The following model is assumed:

$$Y_i = m(X_i) + \sigma(X_i)e_i, i = 1, \ldots, n , \tag{15}$$

where $X_1, \ldots, X_n$ is an iid sample with unknown distribution density f, $e_1, \ldots, e_n$ is an iid sample, independent of the X_i, of random variables with expectation 0 and variance 1. Note $\varepsilon_i = \sigma(X_i)e_i$. We are looking for confidence intervals for $m(x) = E\left(Y \mid X = x\right)$, for some fixed x. A kernel K and a bandwidth h are used to estimate $m(x)$ by

$$\widehat{m}_h(x) = \sum_{i=1}^n \frac{Y_i K\left(\dfrac{x - X_i}{h}\right)}{\sum_{j=1}^n K\left(\dfrac{x - X_j}{h}\right)} = \frac{1}{n} \sum_{i=1}^n W_{hi}(x) Y_i \ .$$

Usual regularity conditions are used to ensure nice asymptotic properties of $\widehat{m}_h(.)$. Details are given in Härdle et al. (1991). In spite of the nonparametric character of the problem, we are faced with two new features: the need to control the bias, and the heteroscedasticity of the errors.

Let's consider both questions in that order. Confidence intervals for $m(x)$ could be based on the fact that, under convenient conditions,

$$S_n = \sqrt{nh} \frac{\widehat{m}_h(x) - m(x)}{V_n^{1/2}} - \frac{B_n}{V_n^{1/2}} \xrightarrow{D} \mathcal{N}(0,1) \ ,$$

where B_n is the bias of $\widehat{m}_h(x)$ and V_n its variance. As it's well known, the size of the bias depends on the tuning of the convergence to 0 of h with respect to the convergence to infinity of n. But S_n isn't practically useful, as B_n and V_n are both unknown. They have to be estimated. A natural estimator $\widehat{V}_n$ of V_n is given by

$$\widehat{V}_n = \frac{h}{n} \sum_{i=1}^n W_{hi}^2(x) \widehat{\varepsilon}_i \ ,$$

with $\widehat{\varepsilon}_i = Y_i - \widehat{m}_h(X_i)$. To compute an estimator of B_n, another estimator of $m(.)$ is needed. We choose

$$\widehat{B}_n = \left(\frac{h}{n}\right)^{1/2} \sum_{i=1}^n W_{hi}(x) \widehat{m}_g(X_i) - \widehat{m}_g(x) \ ,$$

where $\widehat{m}_g$ is a kernel estimate possibly obtained with another kernel, and a bandwidth g. Alternatively, the bandwidth h could be chosen in order to make the bias negligible. This subject isn't treated in the present paper.

The asymptotic pivot T_n is introduced:

$$T_n = \sqrt{nh} \frac{\widehat{m}_h(x) - m(x)}{\widehat{V}_n^{1/2}} - \frac{\widehat{B}_n}{\widehat{V}_n^{1/2}} \ .$$

In the sequel, the bandwidths are chosen as $h = n^{-a}$ and $g = n^{-b}$, a and b being in the interval $]0, 1[$. It can be shown that:

$$\limsup_{n \to \infty} n^{3/8} \left| \Pr\{T_n \leq t\} - \Phi(t) \right| < \infty \ ,$$

if $a = 1/4$ and $b = 1/8$, and that this is the best achievable rate among the consistent choices of a and b. Thus, using T_n, the control of the bias is given a solution.

To overcome the difficulty introduced by the heteroscedasticity of the model, the *wild bootstrap*, as defined by Härdle and Mammen (1991), is used. Roughly speaking, the wild bootstrap enables the resampling of a single observation. More precisely, pseudo-observations Y_i^* are generated according to the scheme

$$Y_i^* = \widehat{m}_{g^*}(X_i) + \varepsilon_i^* \ , \quad i = 1, \ldots, n \ ,$$

where the ε_i^* are independent variables such that $E(\varepsilon_i^*) = 0$, $E(\varepsilon_i^{*\,2}) = \hat{\varepsilon}_i^2$ and $E(\varepsilon_i^{*\,3}) = \hat{\varepsilon}_i^3$. $\widehat{m}_{g^*}(.)$ is a kernel estimate of $m(.)$ with a bandwidth g^*.

The counterpart of T_n in the bootstrap world is

$$T_n^* = \sqrt{nh}\,\frac{\widehat{m}_h^*(x) - \widehat{m}_{g^*}(x)}{\widehat{V}_n^{*\,1/2}} - \frac{\hat{B}_n^*}{\widehat{V}_n^{*\,1/2}} \ ,$$

with

$$
\begin{aligned}
\hat{\varepsilon}_i^* &= Y_i^* - \widehat{m}_h^*(X_i) \ , \\
\widehat{V}_n^* &= \frac{h}{n} \sum_{i=1}^n W_{hi}^2(x)\hat{\varepsilon}_i^* \ , \\
\widehat{m}_g^*(x) &= \frac{1}{n} \sum_{i=1}^n W_{gi}(x)Y_i^* \ , \\
\hat{B}_n^* &= \left(\frac{h}{n}\right)^{1/2} \sum_{i=1}^n W_{hi}(x)\widehat{m}_g^*(X_i) - \widehat{m}_g^*(x) \ .
\end{aligned}
$$

Let $g^* = n^{-c}$. Choosing $a = 2/7$, $b = 1/7$ and $c \in]3/28, 2/7[$, the following approximation is achieved:

$$\limsup_{n \to \infty} \sup_{t \in \mathbf{R}} n^{3/8} \left|\Pr\{T_n \le t\} - \Pr\{T_n^* \le t \,|\,\mathcal{Y}\}\right| = 0 \ , \quad \text{a.s.}$$

Moreover, define $\mathsf{v}_\alpha(\mathcal{Y}) = \inf\{t : \Pr\{T_n^* \le t \,|\,\mathcal{Y}\} \ge \alpha\}$. Then, bootstrap based confidence intervals for $m(x)$ can be computed, achieving a better rate of convergence of their actual level to their nominal value than the one based on the first order asymptotic theory. That is to say

$$\Pr\left\{m(x) \ge \widehat{m}_h(x) - \frac{\hat{B}_n}{\sqrt{nh}} - \frac{\widehat{V}_n^{1/2}}{\sqrt{nh}}\mathsf{v}_\alpha(\mathcal{Y})\right\} = \alpha + \mathrm{O}(n^{-5/7}) \ .$$

7   Concluding remarks

We didn't comment the results displayed on tables (1) to (4). As indicated earlier, they only are part of the outcomes of larger simulation studies. The reported figures, as well as the other results, support the conjecture that, for semiparametric (and also parametric) regression models, the nonparametric bootstrap provides an effective method for computing confidence sets achieving an actual coverage probability not too far from the prescribed one, even in small samples situations. The asymptotic framework, assuming an infinitely increasing number of observations, provides a tool for the comparison of the performance of different confidence intervals computation methods, which doesn't distort too much their actual relative behaviours.

From this last point of view, the situation encountered for the Gaussian analysis of variance model is rather similar. Actually, the number of observations is held fixed, and all the theoretical properties proceed from the Gaussian distribution of the observations. Anyway, in this case, the main theoretically justified outcome is not to improve the approximation of some distribution function by the bootstrap, but to avoid difficult numerical computations. Here too, the practical usefulness of the bootstrap has been shown.

On the other hand, for the nonparametric regression models, only theoretical results have been obtained. They are in agreement with the general *second order correctness* theory of the bootstrap, because the wild bootstrap resampling scheme, in spite of its relative roughness, provides estimates of the first moments of the residuals sufficiently close to their real values. As a matter of fact, the kernel estimation for nonparametric regression models is a computer intensive method by itself. Using the empirical quantiles of T_n^*, $4 + 8B$ kernel smoothings will be needed for computing the bounds of a confidence interval for T_n, where B is the number of bootstrap simulations. Moreover, the choice of the bandwidths in actual cases has not been considered so far. So, if the theory, on one hand, and the successful practical implementation of bootstrap procedures in other situations, on the other hand, lead to conjecture that something could be gained, in terms of accuracy of confidence interval level, by using resampling methods, the price to pay, in terms of computing time, is certainly very high.

References

BERAN, R. (1987). Prepivoting to reduce level error of confidence sets. *Biometrika.* **74** 457–468.

BHATTACHARYA, R. N. and GHOSH, J. K. (1978). On the validity of the formal Edgeworth expansion. *Annals of Statistics.* **6** 434–451.

CHADŒUF, J. and DENIS, J. B. (1988). The analysis of nonadditivity in two-way analysis of variance. *Journal of Applied Statistics.* **18** 331–353.

CSŐRGŐ, S. and TOTIK, V. (1983). On how long interval is the empirical characteristic function uniformly consistent? *Acta Scientifica Mathematica (Szeged).* **45** 141–149.

GOODMAN, L. A. and HABERMAN, S. J. (1990). The analysis of nonadditivity in two-way analysis of variance. *Journal of the American Statistical Association.* **85** 139–145.

GRUET. M. A. and JOLIVET, E. (1991). Calibration with a nonlinear standard curve: asymptotic properties and bootstrap procedures. Technical report. INRA, Département de Biométrie.

HALL, P. (1983). Inverting an Edgeworth expansion. *Annals of Statistics.* **11** 569–576.

HALL, P. (1986). On the bootstrap and confidence intervals. *Annals of Statistics.* **14** 1431–1452.

HALL, P. (1988). Theoretical comparison of bootstrap confidence intervals. *Annals of Statistics.* **16** 927–953.

HÄRDLE, W., HUET, S. and JOLIVET, E. (1991). Better bootstrap confidence intervals for regression curve estimation.

HÄRDLE, W. and MAMMEN, E. (1991). Bootstrap simultaneous errors bars for nonparametric regression. *Annals of Statistics.* ???–???

HUET, S. (1991). Méthodes asymptotiques pour le calcul d'intervalles de confiance en régression non-linéaire, quand la variance tend vers 0. Application aux modèles d'interaction. *Comptes-Rendus de l'Académie des Sciences, Série mathématiques.* 937–942.

HUET, S., JOLIVET, E. and MESSÉAN, A. (1989). Some simulations results about confidence intervals and bootstrap methods in nonlinear regression. *Statistics.* **21** 369–432.

HUET, S. and JOLIVET, E. (1989). Exactitude au second ordre des intervalles de confiance bootstrap pour les paramètres d'un modèle de régression non-linéaire. *Comptes-Rendus de l'Académie des Sciences, Série mathématiques.* 429–432.

JENNRICH, R. (1969). Asymptotic properties of nonlinear least squares estimators. *The Annals of Mathematical Statistics.* **40** 633–643.

SCHMIDT, W. and ZWANZIG, S. (1986). Second order asymptotics in nonlinear regression. *Journal of Multivariate Analysis.* **18** 187–215.

SINGH, K. (1981). On the asymptotic accuracy of Efron's bootstrap. *The Annals of Statistics.* **9** 1187–1195.

Application of Resampling Methods to the Choice of Dimension in Principal Component Analysis

Ph. Besse and A. de Falguerolles

Laboratoire de Statistique et Probabilités, U.A. CNRS D0745, Université Paul Sabatier, 31062 Toulouse cedex, France.

Abstract

This paper investigates the problem of the choice of dimension in Principal Component Analysis (PCA). PCA is introduced as a model; a loss function assessing the stability of the fit is considered. The choice of dimension then amounts to the minimisation of an expected loss which has to be estimated. This is achieved by resampling methods. Different bootstrap and jackknife estimates are presented. The behaviour of these estimates are investigated on artificial data and on real data. The resulting choices are confronted with those given by naïve rules.

Keywords: Principal Component Analysis; Optimal Dimension; Bootstrap; Jackknife; Perturbation Theory.

1   Introduction

Principal Component Analysis (PCA) is usually defined as the search for independent linear combinations of the initial variables with maximal variances. It can be also viewed as a dimension reduction of the data aiming at producing graphical displays in a low dimensional subspace. In accordance with Caussinus (1986), PCA is introduced in this paper as a fixed effect model. This classical modelling approach gives rise to a measure of fit which can be formulated in terms of a risk which is the expectation of a quadratic distance. This risk can then be used in order to optimise the choice of a parameter: a metric to measure distances between rows (see Besse et al. 1988), a smoothing parameter (see Besse and Pousse 1992) or the number of dimensions.

This paper discusses this last situation. In section 2, we introduce the model and the risk function. In this context, the choice of dimension amounts to the minimisation of an expected loss function. This is achieved by resampling methods which are presented in section 3. Bootstrap and jackknife estimates are reviewed. An approximation of the latter which is simply derived from the results of the PCA of the data is presented. In section 4, we illustrate the use of these estimates on artificial and real data. We compare their performances to those of certain naïve decision rules based on the results of a standard PCA. A new rule based on the shape of the parallel boxplots of the principal components is introduced. Finally, in section 5 we make some concluding remarks.

168

2 A PCA stability criterion

2.1 Model and notation

Let $\{Y_i; i = 1, \ldots, n\}$ be n independent vector valued random variables in $\mathbf{R}^p$. The fixed effect model assumes that each realization y_i can be decomposed as a fixed effect $\mu + z_i$ and an error term ε_i:

$$y_i = \mu + z_i + \varepsilon_i \tag{1}$$

$$\text{with } \mu, z_i \in \mathbf{R}^p, \sum_{i=1}^n z_i = 0 \quad and, \quad \forall i, E(\varepsilon_i) = 0, \text{var}(\varepsilon_i) = \sigma^2 \mathbf{I}_p.$$

Furthermore, we assume that vectors $\{z_i; i = 1, \ldots, n\}$ lie in a q dimensional subspace $\mathcal{E}_q$ of $\mathbf{R}^p$.

The least squares estimation of this model is obtained by solving the following minimisation problem:

$$\min_{\mathcal{E}_q, \mu, z_i} \left\{ \frac{1}{n} \sum_{i=1}^n \|y_i - \mu - z_i\|^2 ; \mu \in \mathbf{R}^p, \dim(\mathcal{E}_q) = q, z_i \in \mathcal{E}_q, \sum_{i=1}^n z_i = 0 \right\}. \tag{2}$$

Let $\bar{y} = \frac{1}{n} \sum_{i=1}^n y_i$ be the column mean; then

$$\sum_{i=1}^n \|y_i - \mu - z_i\|^2 = \sum_{i=1}^n \|y_i - \bar{y} - z_i\|^2 + n \|\bar{y} - \mu\|^2$$

Hence, denoting by $\mathbf{X}$ the column centered data matrix and by $\mathbf{Z}$ the (n, p) matrix whose rows are the z_i, the minimization problem (2) is equivalent to:

$$\min_{\mathbf{Z}, \mu} \left\{ \frac{1}{n} \|\mathbf{X} - \mathbf{Z}\|_2^2 + \|\bar{y} - \mu\|^2 ; \mu \in \mathbf{R}^p, \text{rank}(\mathbf{Z}) = q, \sum_{i=1}^n z_i = 0 \right\}, \tag{3}$$

where $\|.\|_2$ is the SSQ norm on (n, p) matrices.

The solution of that problem is well known and is given by considering the order q generalized singular value decomposition (SVD) of $\mathbf{X}$:

$$\widehat{\mathbf{Z}_q} = \sum_{k=1}^q \lambda_k^{1/2} u_k v_k', \tag{4}$$

where u_k and v_k are respectively the orthogonal eigenvectors of $\mathbf{X}\mathbf{X}'/n$ and those of the empirical covariance $\mathbf{S} = \mathbf{X}'\mathbf{X}/n$ —or correlation matrix in case of standardized variables— associated with the eigenvalues λ_k arranged in decreasing order. We denote by $\mathbf{V}_q$ the (p, q) matrix which stores the column vectors $\{v_k; k = 1, \ldots, q\}$ and by $\mathbf{U}_q$ the (n, q) matrix which stores the column vectors $\{u_k; k = 1, \ldots, q\}$. The different estimates in problems (2) and (3) are then given by:

$$\begin{aligned} \widehat{\mu} &= \bar{y}, \\ \widehat{\mathcal{E}_q} &= \text{vect}\{v_k; k = 1, \ldots, q\}, \text{ subspace spanned by the } v_k \text{s}, \\ \widehat{P_q} &= \mathbf{V}_q \mathbf{V}_q', \text{ the eigenprojector matrix}, \\ \widehat{z_i} &= \widehat{P_q} x_i, \\ \widehat{\mathbf{Z}_q} &= \mathbf{X}\widehat{P_q}. \end{aligned}$$

Moreover,

$$\mathbf{C}_q = \mathbf{U}_q \text{diag}(\lambda_k; k = 1, \ldots, q) = \mathbf{X}\mathbf{V}_q$$

gives the q principal variables or component scores which lead to graphical displays for the rows.

In fact, the quality of all these estimates seriously depends on the choice of the number q of dimensions or components which are retained. Our aim is to define an approximation which should remain stable with respect to sampling alterations. Therefore a criterion of stability has to be defined.

2.2 Stability criterion

A natural criterion for the stability of the vectors z_i (see Besse et al. 1988) is given by the following loss function:

$$L_q = \frac{1}{n} \sum_{i=1}^{n} \|z_i - \widehat{z}_i\|_\mathbf{A}^2 \,,$$

where $\mathbf{A}$ defines any euclidean metric in $\mathbf{R}^p$. Its expectation, the associated risk function, is used to optimize the euclidean metric within the vector space $\mathbf{R}^p$; it leads to a Gauss-Markov-like property in the framework of PCA: the optimal metric is given by the inverse matrix of the covariance of the noise. When estimating the dimension, this approach unfortunately raises a problem of indetermination: it is not possible to estimate simultaneously σ and the optimal dimension q because the natural estimate for σ^2 is the average of the $(p-q)$ smallest eigenvalues.

This drawback can be overcome by considering the subspace $\widehat{\mathcal{E}}_q$, which is asymptotically consistant (see Fine and Pousse 1990) rather than the vectors z_i. The loss function then becomes:

$$L_q = Q(\mathcal{E}_q, \widehat{\mathcal{E}}_q) = \frac{1}{2} \left\| P_q - \widehat{P}_q \right\|_2^2 = q - \mathrm{tr} P_q \widehat{P}_q, \tag{5}$$

where Q is based on the usual SSQ norm of matrices which is applied to measure distances between projectors and thus to measure distances between the associated subspaces. In that context, $\mathrm{tr} P_q \widehat{P}_q$ is also the sum of the squared canonical correlation coefficients between the component sets which respectively span $\mathcal{E}_q$ and $\widehat{\mathcal{E}}_q$. This approach follows the principle which underlies indices based on the sum of correlation coefficients between pairs of principal components which have been introduced by Daudin *et al.* (1988,1989): in a PCA, the results are assumed to be reliable if the representation subspace is stable.

Finally, the risk function is defined by considering the following expectation:

$$R_q = E Q(\mathcal{E}_q, \widehat{\mathcal{E}}_q). \tag{6}$$

To compute estimates for this risk function, we resort to computer intensive methods. R_q being symmetrically defined since its value is invariant under any permutation of the observations y_i, resampling methods such as the bootstrap and the jackknife are natural candidates. This strategy has been previously advocated in a different context by Beran and Srivastava (1985) in order to estimate confidence regions for differentiable functions of the covariance matrix or by Daudin et al. (1988, 1989) to study certain stability indices of PCA.

3 Risk estimates

3.1 Bootstrap

According to Efron (1982), a bootstrap estimate of a mean square risk is defined by:

$$\widehat{R_{Bq}} = \frac{1}{B} \sum_{b=1}^{B} (q - \mathrm{tr} P_q^{*b} \widehat{P}_q) = q - \frac{1}{B} \sum_{b=1}^{B} \mathrm{tr} P_q^{*b} \widehat{P}_q, \tag{7}$$

where B is the number of bootstrap replications, P_q^{*b} is the eigenprojector which is obtained by computing the bth sample PCA.

3.1.1 Bootstrap of residuals

In model (1) the errors ε_i are assumed to be i.i.d. Thus the most natural way to compute an estimate for R_q is to bootstrap the residuals. This leads to the following algorithm[1]:

---

[1] all algorithms have been programmed in S language (see Becker et al. 1988).

```
Compute the PCA of X
For q = 1 to (p − 1) do
```

$\qquad$ Calculate residuals: $\quad e_i = \lambda_{q+1}^{1/2} \mathbf{V} u_i; i = 1, \ldots, n$
$\qquad$ whose covariance matrix is $\lambda_{q+1} \mathbf{I}_p$

$\qquad$ **Repeat** B **times**

$\qquad\qquad$ Draw the e_i^* with replacement from the e_i
$\qquad\qquad$ Construct a bootstrap sample $\mathbf{X}^*$: $\quad x_i^* = \widehat{P_q} x_i + e_i^*$
$\qquad\qquad$ Compute the PCA of $\mathbf{X}^*$
$\qquad\qquad$ Cumulate: $\operatorname{tr}\widehat{P_q} P_q^{*b}$

$\qquad$ **End**
$\qquad$ Calculate: $\widehat{R_{Brq}}$ by (7)

End
Compare $\widehat{R_{Brq}}$ values

Unfortunately, this requires a large amount of computation: B times $(p-1)$ singular value decompositions (SVD) must be performed. Indeed this can be computationaly expensive and time consuming. Hence we propose a moderate value for B, say 30 .

3.1.2 Bootstrap of data

A less expensive estimate can be computed by naïvely bootstrapping the data. It should be emphasized that this is somewhat contradictory with model (1) since a bootstrap of the data assumes that the Y_i are i.i.d.. The algorithm becomes:

```
Compute the PCA of X
Repeat B times
```

$\qquad$ Construct $\mathbf{X}^*$ by drawing the x_i^* with replacement from the x_i
$\qquad$ Compute the PCA of $\mathbf{X}^*$
$\qquad$ **For** $q = 1$ **to** $(p-1)$ **do**
$\qquad\qquad$ Cumulate: $\operatorname{tr}\widehat{P_q} P_q^{*b}$
$\qquad$ **End**

End
Calculate: $\widehat{R_{Bdq}}$ by (7)
Compare $\widehat{R_{Bdq}}$ values

The algorithm performs B times a SVD to compute the naïve bootstrap estimate. Hence we propose the value $B = 100$.

3.2 Jackknife estimate

Under the same rough assumptions as those considered for the bootstrap of the data, we can also compute the jackknife estimate of R_q. The jackknife estimate (see Efron 1982) is defined by:

$$\widehat{R_{JKq}} = (n-1)\left(q - \frac{1}{n}\sum_{i=1}^{n} \operatorname{tr} P_q^{(i)} \widehat{P_q}\right). \tag{8}$$

where we denote by $P_q^{(i)}$ the eigenprojector obtained by performing a PCA of the data matrix $\mathbf{X}^{(i)}$ obtained from $\mathbf{X}$ by deleting the ith row.

This estimate is obtained by the following algorithm:

```
Compute the PCA of X
For i = 1 to n

        Compute the PCA of X^(i)
        For q = 1 to (p − 1) do
            Cumulate:  tr P_q^(i) P̂_q
    End

    End
    Calculate:  R̂_{JKq} by (8)
    Compare  R̂_q values
```

This risk estimate requires the computation of n SVDs but, if n is too large, the jackknife replications can be performed only $B < n$ times for a random sample of rows and thus with the same computational cost as that of bootstrap. Moreover by using the so-called jackknife-after-bootstrap (Efron, 1992), an approximation of the jackknife can be derived from the bootstrap of the data with supplementary data management: each bootstrap sample gives an approximation of the estimate jackknifed with respect to all units which are not sampled.

3.3   Approximation of the jackknife estimate

All estimates above are computationaly expensive. Hence we now consider an approximation of the jackknife which does not require much computational effort. The elimination of the ith row in the data set $\mathbf{X}$ provides an empirical covariance matrix $\mathbf{S}^{(i)}$ which can be decomposed as follows:

$$\mathbf{S}^{(i)} = \mathbf{S} + (n-1)^{-1}(\mathbf{S} - x_i x_i') - (n-1)^{-2} x_i x_i'. \qquad (9)$$

If n is large enough, it is quite acceptable to consider that any row elimination introduces only a small perturbation in further computations. Perturbation theory (Kato, 1966) then enables us to write Taylor's expansions of the eigenelements of $\mathbf{S}$ and thus of the eigenprojectors which are introduced in (8). These results lead to the following Taylor's expansion of the jackknife estimate (see Besse, 1992 for more details):

$$\widehat{R_{JKq}} = \widehat{R_{Pq}} + O((n-1)^{-2}).$$

$\widehat{R_{Pq}}$, an analytic approximation of the jackknife estimate is given by:

$$\widehat{R_{Pq}} = \frac{1}{n-1} \sum_{k=1}^{q} \sum_{j=k+1}^{p} \frac{\tau_{jk}}{(\lambda_j - \lambda_k)^2} \text{ with } \tau_{jk} = \frac{1}{n} \sum_{i=1}^{n} c_{ik}^2 c_{ij}^2 \qquad (10)$$

where c_{ij} denotes the general entry of the principal components matrix $\mathbf{C}$. It is clear that this approximation is easily computed from the results of the PCA of the initial data set.

The validity condition for this expansion is easy to check. It relates n and q as follows:

$$n > \frac{\|\mathbf{S}\|_2^2}{\inf\{(\lambda_k - \lambda_{k+1}); k = 1, \ldots, q\}}.$$

The results above show the importance of the gap between successive eigenvalues. The leading term in $\widehat{R_{Pq}}$ depends on the difference between the eigenvalues associated with the last retained dimension and the first neglected one. This is consistant with intuitive considerations: if the difference $(\lambda_q - \lambda_{q+1})$ is large enough, data perturbations cannot lead to the swapping of the associated eigenvectors v_q and v_{q+1}. Formula (10) also highlights the influence of the fourth order moments of the principal components: high values for the average τ_{jk} of cross products of squared principal components increase the estimated risk.

Finally this approximation simply extends to the PCA of standardized variables. Indeed, perturbation of the correlation matrix instead of the covariance matrix (9) introduces extra terms, but Besse (1992) shows that they all vanish in the approximation calculus. Thus, Taylor's expansion of the risk estimate (10) also holds for centered and standardized PCA.

Note that Daudin et al. (1989) have also suggested in a similar context the use of the infinitesimal jackknife to avoid large computational costs and to obtain analytic expansions. Unfortunately, in their approach, it is impossible to check the validity of the results since functions of ordered eigenvalues do not have continuous derivatives.

4 Illustration

4.1 Motivation

We now turn our attention to data and present the results obtained in considering artificial and real data. The aim of this illustration is to investigate the behaviour of the different estimates of the risk function and to confront the resulting decisions with those of standard naïve rules. The naïve decision rules which we use as benchmarks are:

- the screegraph (search for a knee),

- the parallel boxplots of the principal components (search for a reduction of the boxes visualising the quartiles of principal components),

- Kaiser's rule as implemented in SAS (1989) automatic selection procedure (selection of eigenvalues larger than average).

For an extensive discussion of such rules and some others, a good reference is Jolliffe (1986).

4.2 Artificial data

The 5 artificial data sets each consist of a $(100, 10)$ matrix: $p = 10$ variables observed on $n = 100$ units. They are obtained by adding to the same matrix of fixed effects with known rank $(q = 4)$ a simulated white noise of controlled standard deviation:

1. The fixed effects have been constructed as follows:

   - first, a $(100, 10)$ matrix of pseudo-random numbers uniformly and independently distributed in the range $[0, 20]$ was generated;

   - then the $(100, 10)$ matrix of non centered fixed effects was derived from its SVD by constructing its best least squares approximation of rank $q = 4$.

2. The noise was simulated as follows:

   - 1000 observations of independently and identically normally distributed pseudo-random variables with mean 0 and variance 1 were generated;

   - 5 simulated noises of standard deviation σ $(\sigma = 1, 3, 5, 7, 10)$ were then obtained by multiplication by σ.

Since the data were simulated, the true quadratic risk can be computed and reported. To mimic practical situations, in the formula giving the quadratic loss L_q we have used the order q approximations of the vectors z_i which coincide with the z_i for q greater or equal to the true

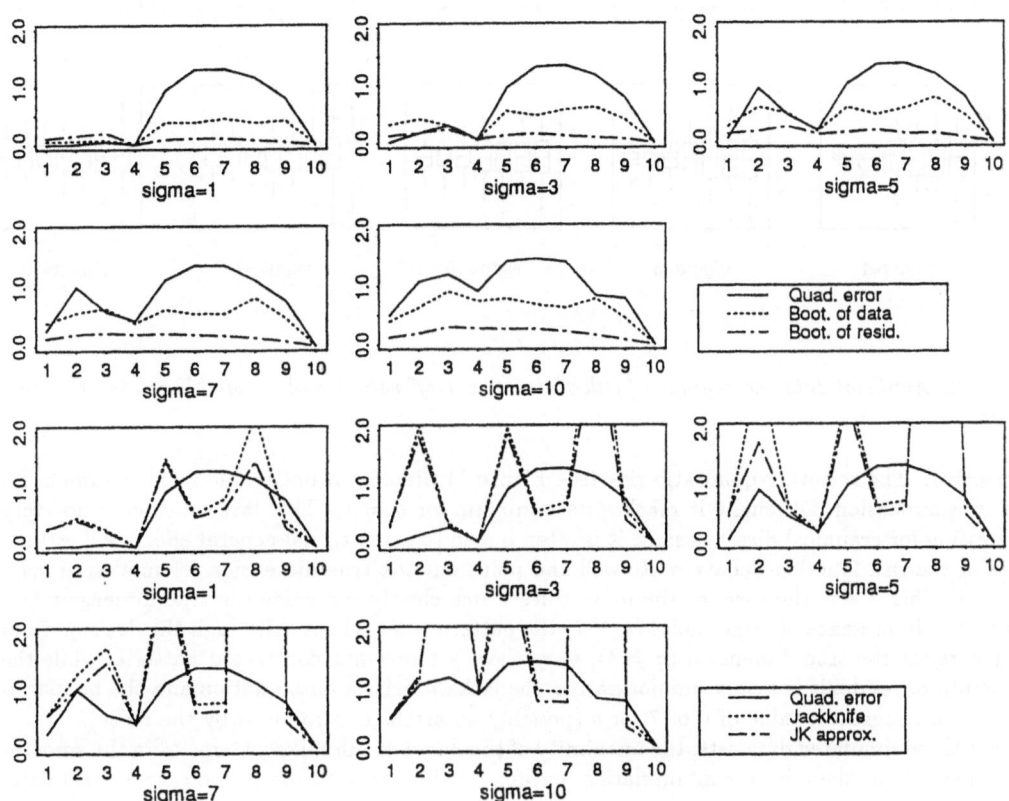

Figure 1: *Artificial data: risk estimates.*

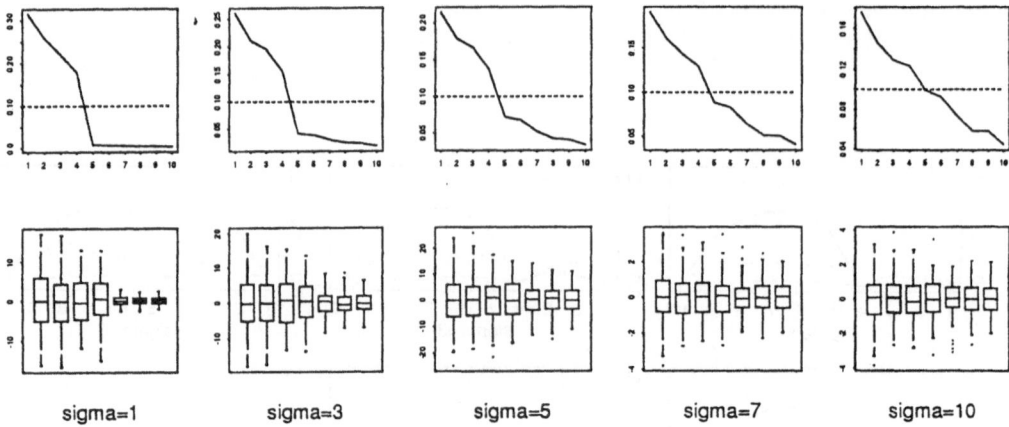

| sigma=1 | sigma=3 | sigma=5 | sigma=7 | sigma=10 |

Figure 2: *Artificial data: screegraphs (with Kaiser's cut-off value) and parallel boxplots of components.*

dimension. The resulting quadratic risk (see Figure 1) presents a noticeable local minimum for the true dimension, althought it reaches its minimum for $q = 1$. This last solution is not very interesting for graphical displays since it is often linked to some trivial general effect. All estimations of the risk function behave quite well and point out the true dimension for moderate noise ($\sigma \leq 5$). This is also the case for the naïve rules which clearly recognize the true dimension (see Figure 2). In presence of larger noise ($\sigma \geq 7$) the pattern is less clear. Although the decision rules do not reject the true dimension ($q = 4$), they show a somewhat contrasted pattern: while the bootstrap of residuals is rather uninformative, the jackknife, its approximation and the bootstrap of data also suggest a value of 6 or 7 for q (possibly an artefact introduced by the noise).

For these simulated data sets, the unanimity of the decision rules likely stems from the smoothness of the data: there is no contamination, outlier,... which is not usually the case for real data.

4.3 Pollution data

We now consider the data that McDonald and Schwing (1973) assembled to study the effects of air pollution on mortality. The $p = 16$ variables were measured at each of $n = 60$ areas in the United States. Our outlook being exploratory, we fit a (standardized) PCA to all variables after a logarithmic transform of the 3 pollution potential variables.

Kaiser's rule (see figure 3) selects $q = 5$ whereas the jump in the screegraph and the size reduction in the parallel boxplots only suggest $q = 3$. Interestingly, the parallel boxplots of the principal components provide some insight into the stability of the first $q \leq 3$ dimensional principal subspaces: moderate box but heavy whiskers of extreme data for the first component, large boxes for the second and the third; smaller boxes with influential points in their whiskers for the last $(p - q)$ components. This is supported by the risk function estimated by bootstrapping or jackknifing the data: the corresponding estimates also point out the stability of the first three-dimensional subspaces. It appears that Kaiser's rule overestimates the dimension needed to obtain a stable principal subspace.

175

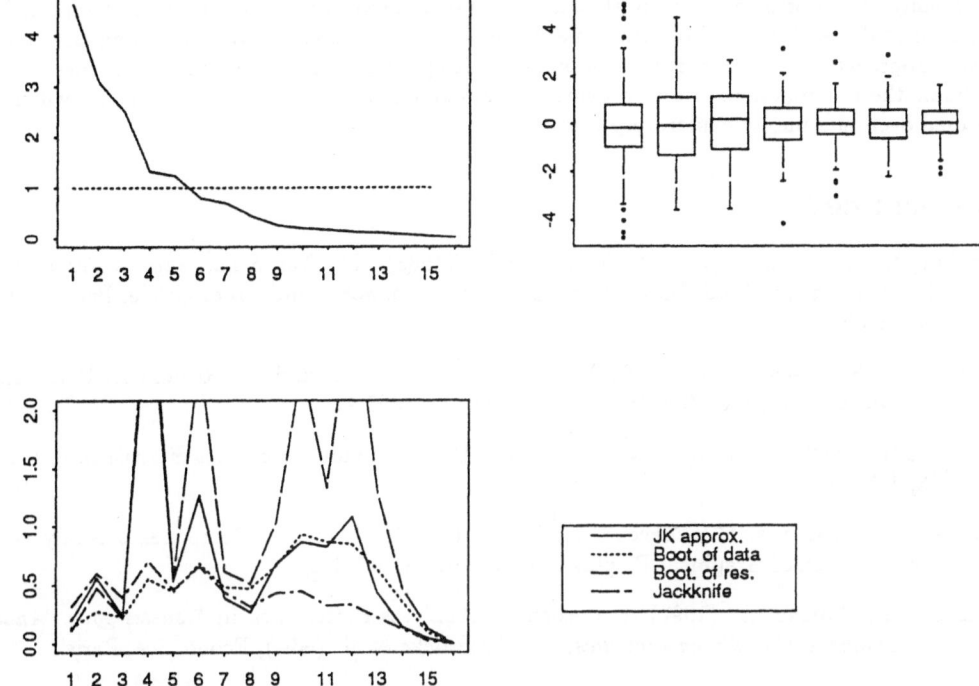

Figure 3: *Pollution data: screegraph (with Kaiser's cut-off rule), parallel boxplots and risk estimates.*

5   Concluding remarks

In the light of our experience of PCA, we recommend the following strategy.

1. For starters, a classical PCA is to be considered. Without much computational effort and therefore on a routine basis, its output can include:

   - the screegraph,
   - the parallel boxplots of the components,
   - the approximation of the jackknife estimate along with its validity conditions relating q and n.

2. These intermediate results enable the stability of the embedded principal subspaces to be assessed. In this respect, we find the parallel boxplots and the approximation of the jackknife very helpful. The former highlights the potential effects of some extreme points while the latter stresses the importance of the gap between the last retained and the first neglected eigenvalues.

3. In case of doubt, a bootstrap approach may provide further insight. When the signal-to-noise ratio is moderate, the bootstrap of residuals should be prefered to the bootstrap of data since it takes advantage of the modelling framework which we are considering.

As seen in the examples, the decision rules may offer several values for the estimated rank q. In this case, we advocate the choice of a low value for the sake of stability. It is clear that the simulated data support our recommendation.

Finally, it is worthwhile noting that the general framework in which we have formulated the problem of dimension selection has some pedagogical implications. Our modelling and optimisation approach gives either natural grounds or insight into the resampling techniques used to estimate the risk function. It also provides useful guidelines for the selection problem, which can be used even in a purely exploratory context.

References

Becker, R.A., Chambers, J.M., Wilks, A.R. (1988). *The New S Language, a Programming Environment for Data Analysis and Graphics*, Wadsworth and Brooks/Cole, Pacific Grove, Ca 93950.

Beran, R., Srivastava, M.S. (1985). Bootstrap Tests and Confidence Regions for Functions of a Covariance Matrix. *The Annals of Statistics*, 13, 95-115.

Besse, Ph. (1992). PCA Stability and Choice of Dimensionality. *Statistics & Probability Letters*, 13, 405-410.

Besse, P., Caussinus, H., Ferré, L., Fine, J. (1988). Principal Components Analysis and Optimization of Graphical Displays. *Statistics*, 19, 301-312.

Besse, Ph., Pousse, A. (1992). Extension des Analyses Factorielles, in *Modèles pour l'Analyse des Données Multidimensionnelles*, J.J. Droesbeke et al. (eds.), Economica, Paris.

Caussinus, H. (1986). Models and Uses of Principal Component Analysis, in *Multidimensional Data Analysis*, J. de Leeuw et al. (eds.), DSWO Press, Leiden, 149-170.

Daudin, J.J., Duby, C., Trécourt, P. (1988). Stability of Principal Component Analysis Studied by Bootstrap, *Statistics*, 19, 241-158.

Daudin, J.J., Duby, C., Trécourt, P. (1989). PCA Stability Studied by the Bootstrap and the Infinitesimal Jackknife Method, *Statistics*, 20, 255-270.

Efron, B. (1982). *The Jackknife, the Bootstrap and other Resampling Methods*, SIAM, Philadelphie.

Efron, B. (1992). Jackknife-after-Bootstrap Standard Errors and Influence Functions (with discussion), *Journal of the Royal Statistical Society*, series B, 54, 83-127.

Fine, J., Pousse, A. (1991). Asymptotic Study of the Multivariate Functional Model; Application to the Metric Choice in PCA, *Statistics*, to appear.

Jolliffe, I. (1986). *Principal Component Analysis*, Springer-Verlag, New-York.

Kato, T. (1966). *Perturbation Theory for Linear Operator*, Springer-Verlag, New-York.

McDonald, G.C., Schwing, R.C. (1973). Instabilities of Regression Estimates Relating Air Pollution to Mortality. *Technometrics*, 15, 463-481.

SAS (1989), SAS/STAT User's Guide, volume 2, Version 6, fourth edition, Sas Institute Inc, Cary.